Materials

A VISUAL APPROACH

Ted Lister
Janet Renshaw

Hodder & Stoughton

A MEMBER OF THE HODDER HEADLINE GROUP

Picture acknowledgements

The publishers would like to thank the following individuals, institutions and companies for permission to reproduce photographs in this book. Every effort has been made to trace ownership of copyright. The publishers would be happy to make arrangements with any copyright holder whom it has not been possible to contact:

Andrew Lambert (64); GeoScience Features Picture Library (98 top, 101); Life File (117, 166 left, 181); Ruth Hughes (175); Science Photo Library (95)/ Bill Bachman (141)/ Alfred Pasieka (98 bottom)/ Lawrence Migdale (126)/ Magrath Photography (166 right)/ Philippe Plailly (14)

Orders: please contact Bookpoint Ltd, 130 Milton Park, Abingdon, Oxon OX14 4SB. Telephone: (44) 01235 827720. Fax: (44) 01235 400454. Lines are open from 9.00 – 6.00, Monday to Saturday, with a 24 hour message answering service. Email address: orders@bookpoint.co.uk

British Library Cataloguing in Publication Data
A catalogue record for this title is available from the British Library

ISBN 0 340 77294 8

First published 2001
Impression number 10 9 8 7 6 5 4 3 2 1
Year 2007 2006 2005 2004 2003 2002 2001

Cover photo from Photodisc
Typeset by Servis Filmsetting Ltd
Printed in Spain for Hodder & Stoughton Educational, a division of Hodder Headline Ltd, 338 Euston Road, London NW1 3BH

Contents

1 Solids, liquids and gases

The three states of matter

the water is a liquid
this rock is a solid
the bubbles of air from the fish are a gas
all the materials around us are either solids, liquids or gases

Solids, liquids and gases are the **three states of matter**.

The differences between solids, liquids and gases

Ice, water and steam are all made from water.

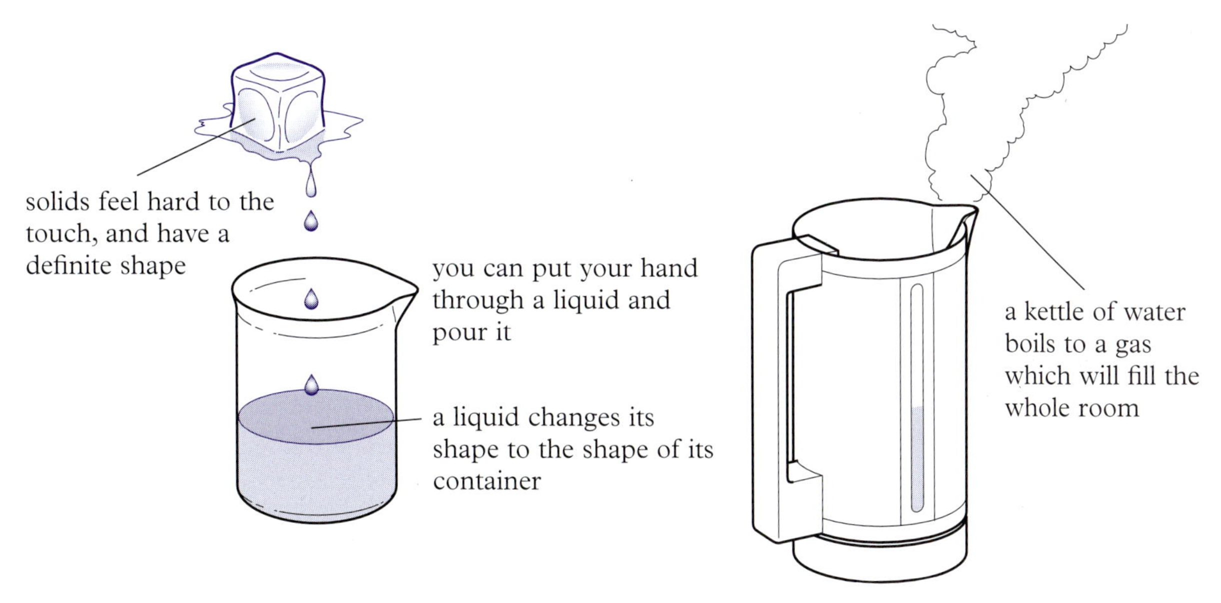
solids feel hard to the touch, and have a definite shape
you can put your hand through a liquid and pour it
a liquid changes its shape to the shape of its container
a kettle of water boils to a gas which will fill the whole room

Particles

All matter is made up of very tiny **moving particles**. These particles are too small to see. It is the differences in the way that these particles behave that makes something a solid, a liquid or a gas. These differences are due to:

- how the particles are arranged – close together or far apart, in any kind of pattern
- how the particles move

How the particles in solids, liquids and gases are arranged

solids have a fixed shape because their particles stay in the same positions

solids have particles which are very close together so you can't squash them into a smaller volume

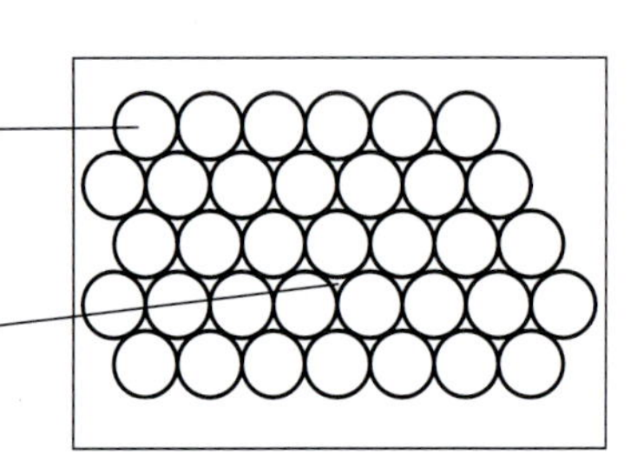

particles in a solid are **regularly arranged** and **close together**

liquids change their shape depending on their container, because their particles have no fixed position

liquids have particles which are close together so you can't squash them into a smaller volume

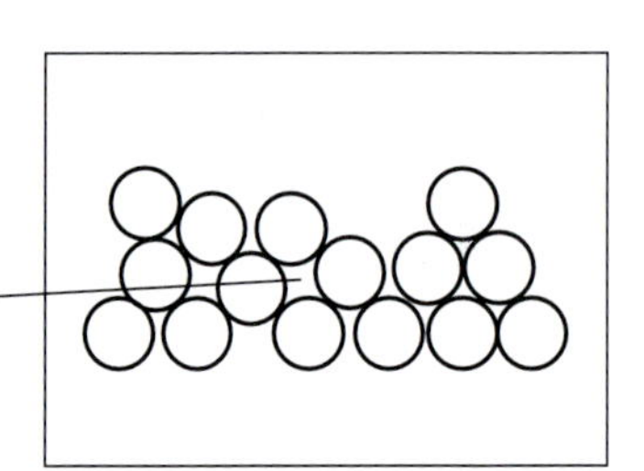

particles in a liquid are **randomly arranged** and **close together**

gases take on the shape of their container, because the particles have no fixed position

gases have a lot of space between the particles so you can squash a gas into a smaller volume

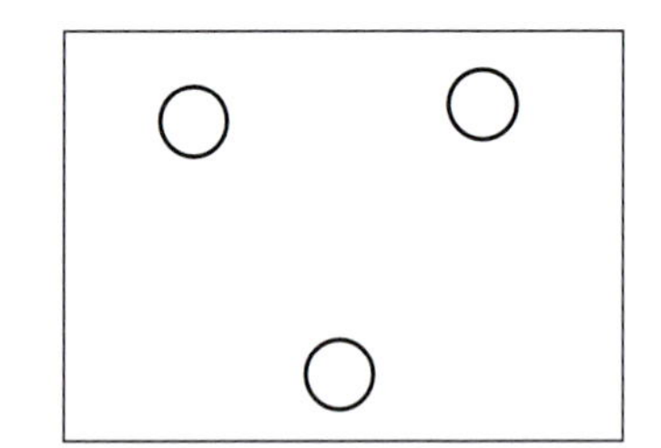

particles in a gas are **randomly arranged** and **far apart**

Questions

1 This question is about how particles are arranged.

Copy the sentences below, filling in the missing words from the following list: **far apart**, **randomly**, **close together**, **fixed positions**.

The same words may be used more than once.

a) The particles in both solids and liquids are c_________ t_________.

b) Particles in solids have f_________ p_________, but those in a liquid do not.

c) The particles in both gases and liquids are r_________ arranged.

d) The particles in a gas are f_________ a_________ but those in a liquid are c_________ t_________.

How the particles in solids, liquids and gases move

Solid

the particles are vibrating around a **fixed** position

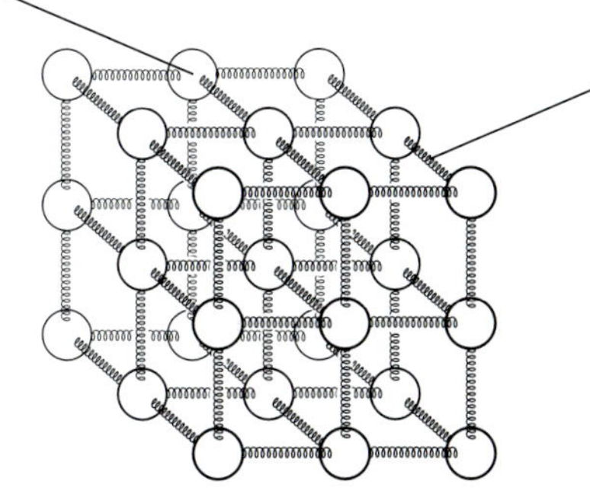

imagine that they are held together by strong springs

solids keep their shape

the particles are strongly attracted to each other

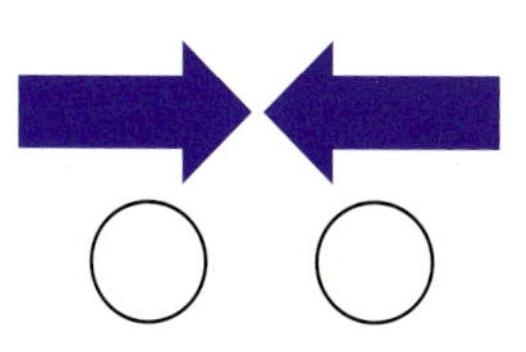

Liquid

the particles are moving rapidly but are close together

they are so close together that they can't move very easily or quickly away from one another

the particles are attracted to each other

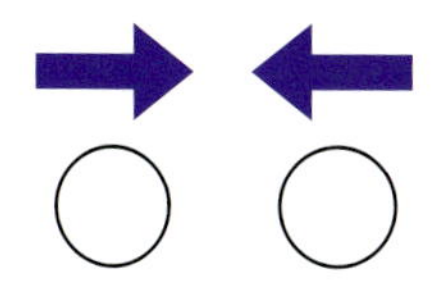

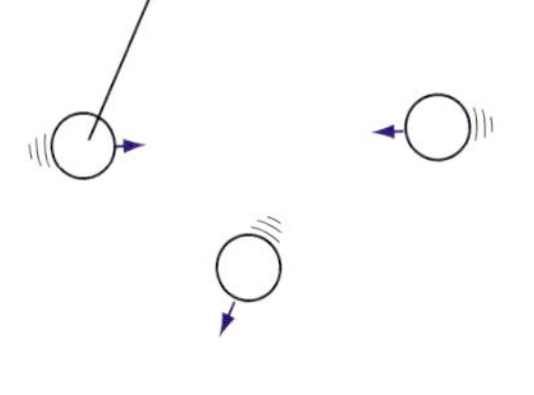

you can pour a liquid from one container to another

Gas

the particles are moving rapidly in all directions

the particles are not close together

the particles are not attracted to each other

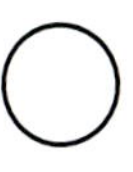

gases always completely fill the container they are in

Questions

1 This question is about how particles move. Copy the following sentences and fill in the missing words from the following list:

far apart, vibrating, moving, close together

The same words may be used more than once.

a) The particles in both solids and liquids stay c________ t________ .

b) The particles in solids are v________ but those in a liquid are m________ from place to place.

c) The particles in both gases and liquids are m________ rapidly.

d) The particles in a liquid are c________ t________ but those in a gas are f________ a________ .

Expansion

When you heat solids, liquids or gases they all take up more space. This means they **expand**. The particles themselves don't get any bigger but they do get further apart.

Solid

Liquid

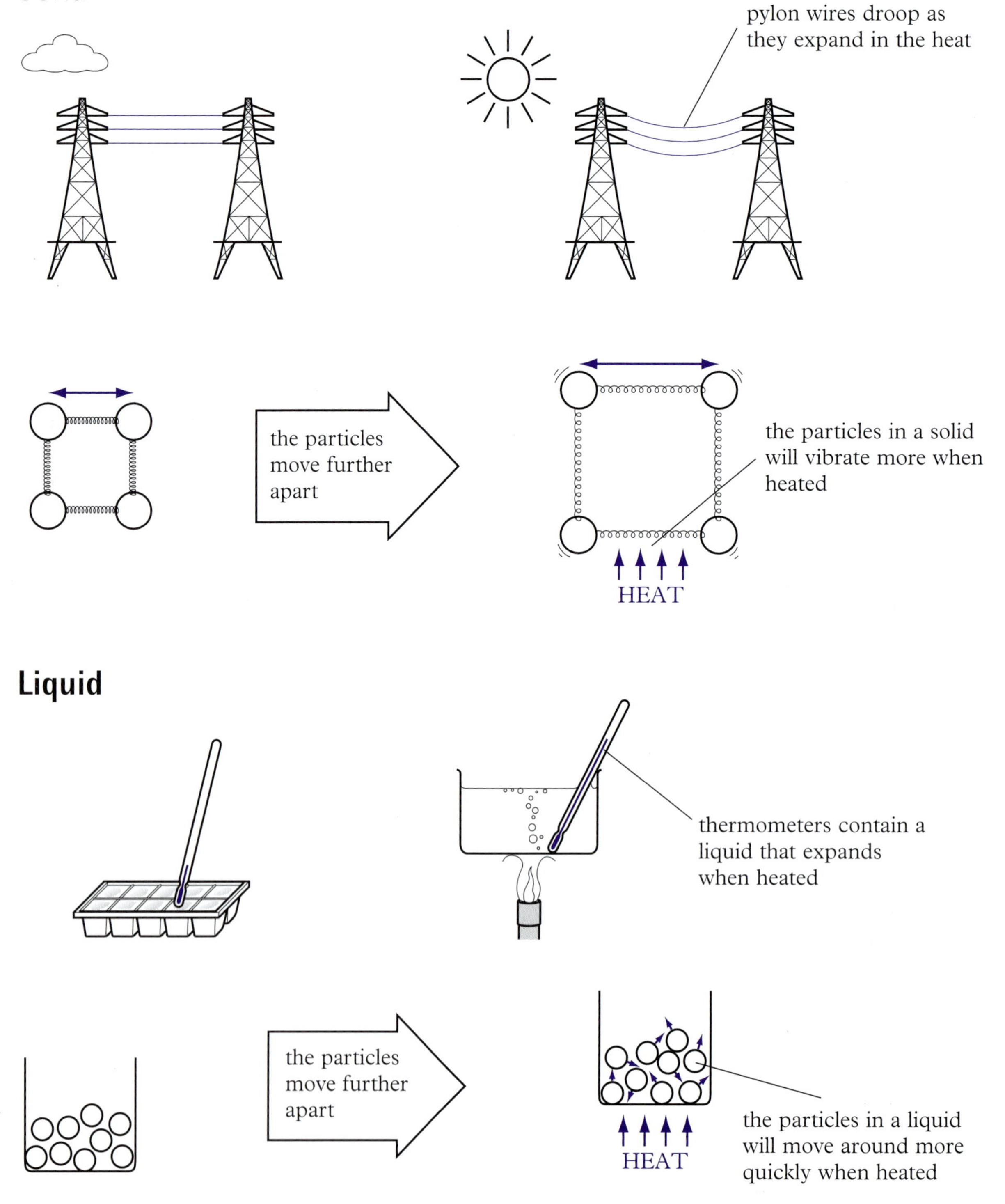

pylon wires droop as they expand in the heat
the particles move further apart
the particles in a solid will vibrate more when heated
HEAT
thermometers contain a liquid that expands when heated
the particles move further apart
HEAT
the particles in a liquid will move around more quickly when heated

Gas

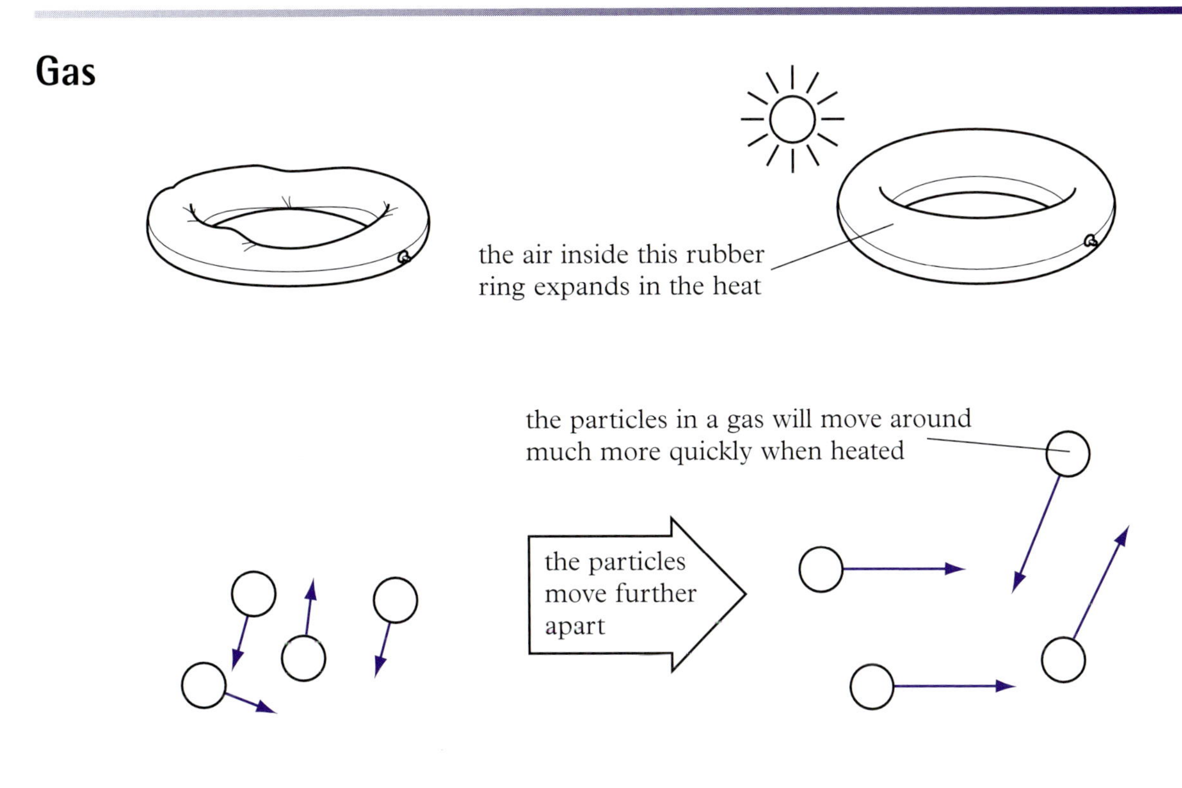
the air inside this rubber ring expands in the heat
the particles in a gas will move around much more quickly when heated
the particles move further apart

Questions

1 Air in a syringe can be squeezed into a smaller space. Water in a syringe cannot be squeezed into a smaller space.

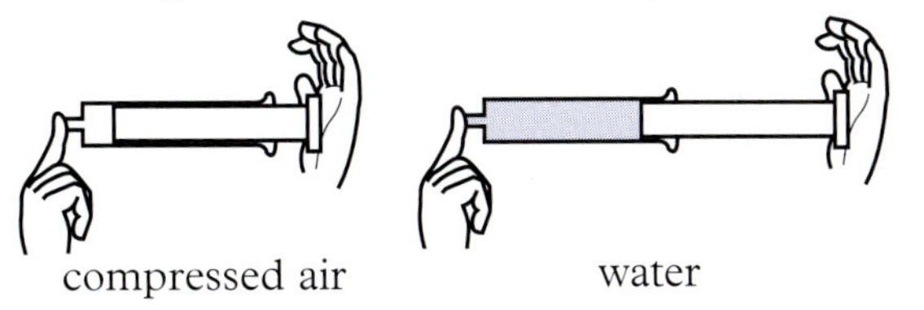
compressed air
water

Explain why squeezing air and water give different results.

2 Use the information about particles to explain the following observations:
a) solids feel hard
b) you can put your hand right into a liquid
c) you do not notice the air around you as you walk through it.

3 Copy and complete the sentence with one of the phrases below:

get bigger get smaller stay the same size

Particles_________ when a solid, a liquid or a gas expands.

4 Explain why solids, liquids and gases expand when you heat them.

Diffusion

Particles are always moving. If you leave them, they spread out and mix together. This is called **diffusion**.

Diffusion of gases

When the doors in a house are left open, the smell of food can pass from room to room by diffusion. If the doors are closed they form a barrier which stops the particles spreading from one room to another.

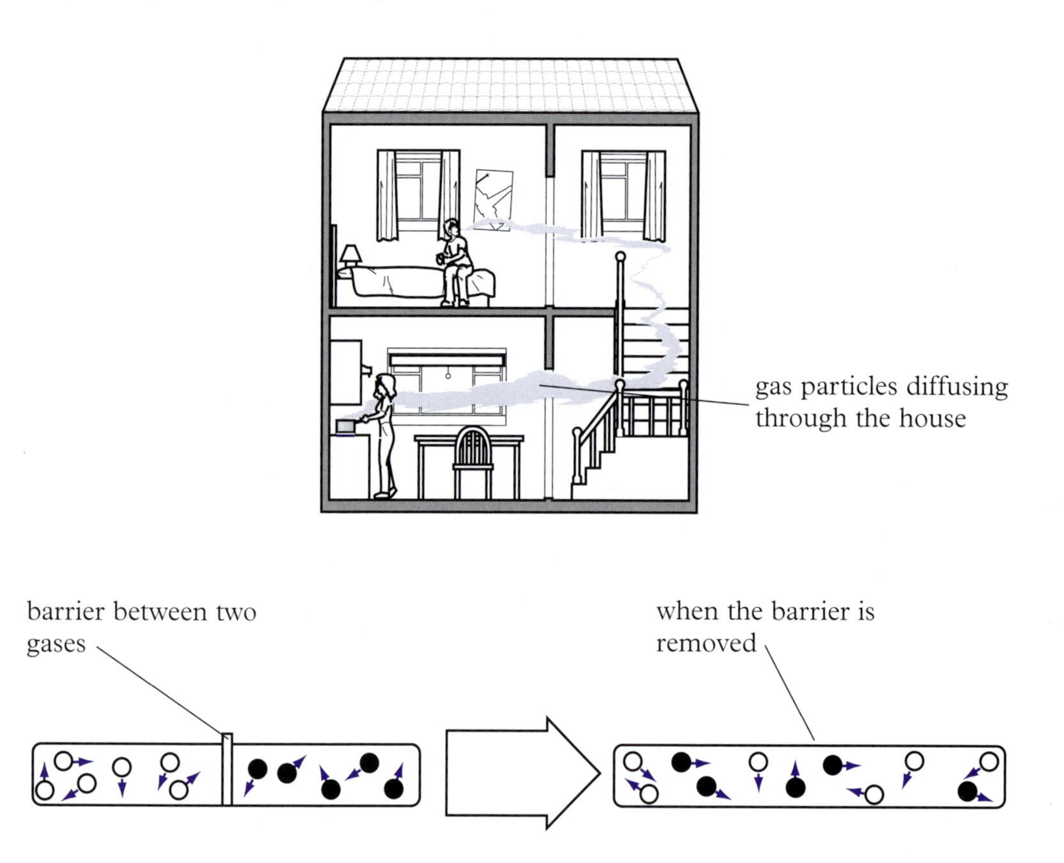
gas particles diffusing through the house
barrier between two gases
when the barrier is removed

Because their particles are moving all the time, gases will spread out and rapidly fill any space that they are in. Two gases will very quickly mix together all by themselves.

Diffusion of liquids

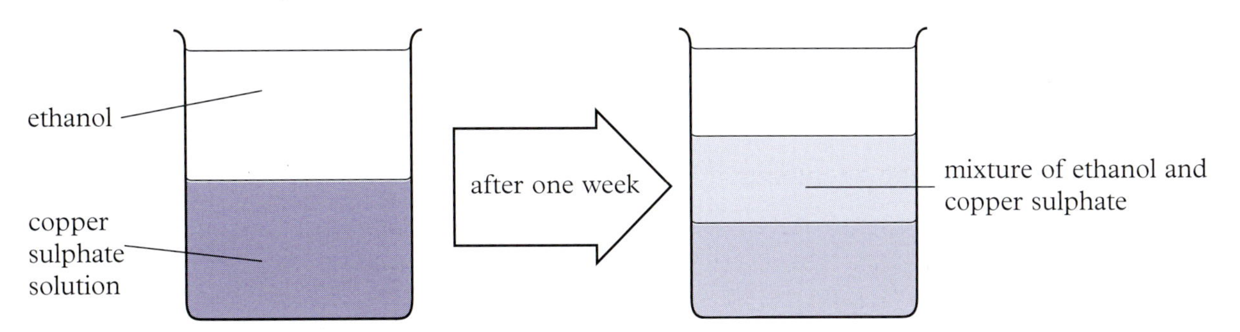
ethanol
copper sulphate solution
after one week
mixture of ethanol and copper sulphate

The particles of a liquid are always moving. Liquids always mix because they will very slowly diffuse into each other all by themselves. Diffusion will be faster if the liquids are warmed.

Pressure

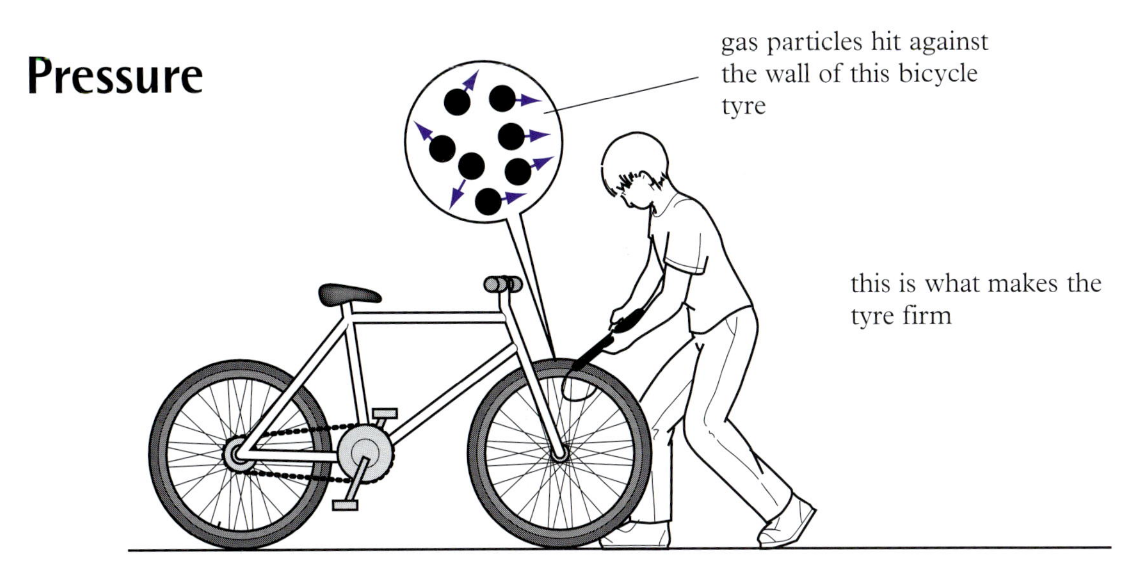
gas particles hit against the wall of this bicycle tyre
this is what makes the tyre firm

Gas particles pound against the walls of their container and will exert a **pressure**. The gas particles move faster when the air inside the tyre is heated. This means the particles hit the walls of the tyre harder.

Questions

1 Use words or labelled diagrams to explain:

a) how a tiny purple crystal is dissolved by water particles

b) why the purple colour of the crystal slowly spreads out (diffuses) throughout the water

c) why the purple colour of the crystal would spread out faster in hot water.

2 The air in a bicycle tyre heats up when a cyclist rides along a road.

a) Will the pressure in a bicycle tyre be the same at the beginning and the end of a long cycle ride?

a) Explain your answer.

3 If a firm rubber ring is taken from a hot sunny beach into the sea it goes softer. Explain why the rubber ring feels softer in the sea.

Changing from one state to another

The three states of matter are solid, liquid and gas. Energy is needed to change matter from one state to another. This energy is provided in the form of heat.

By heating we can change a solid into a liquid and a liquid into a gas.

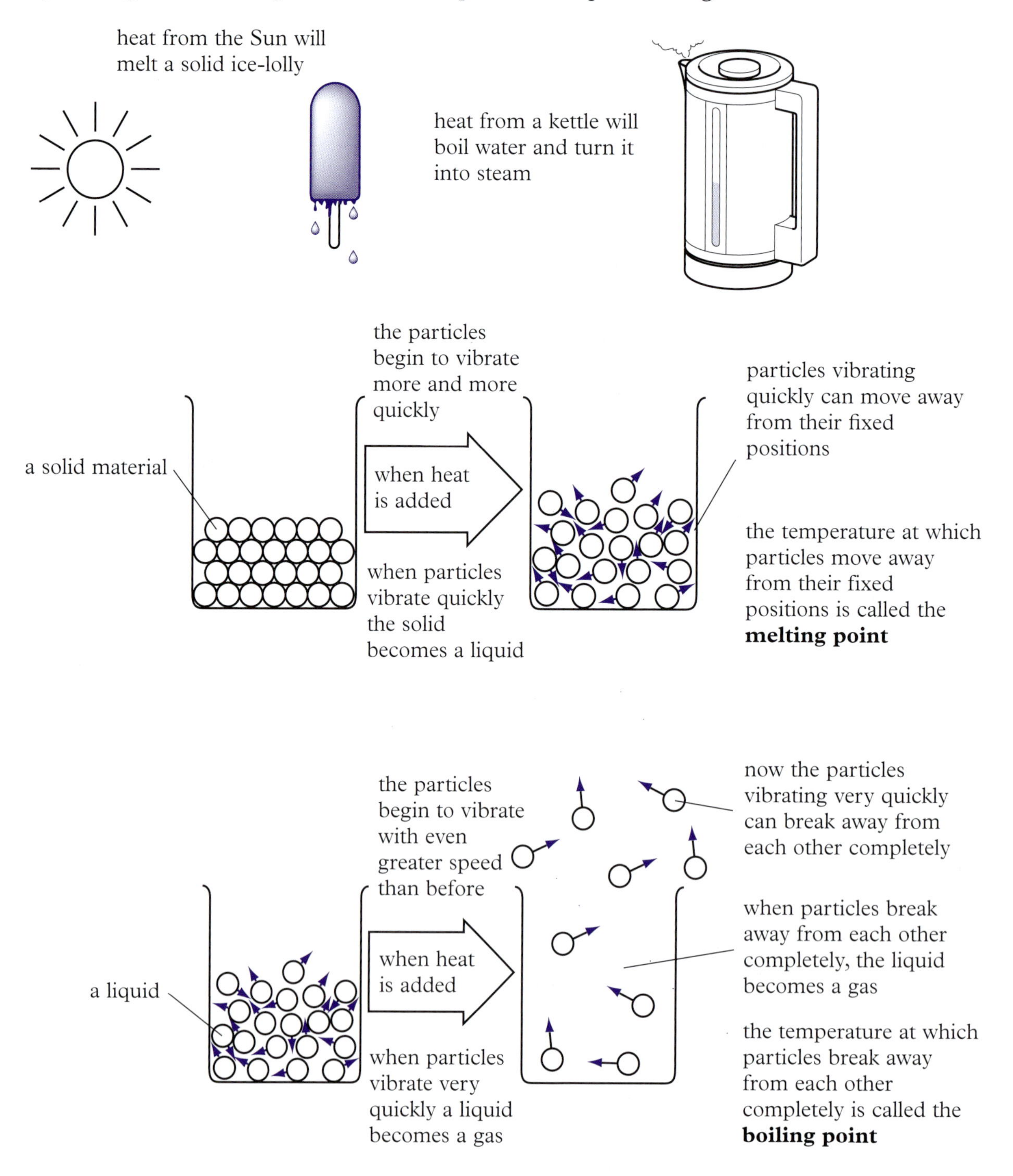
heat from the Sun will melt a solid ice-lolly
heat from a kettle will boil water and turn it into steam
the particles begin to vibrate more and more quickly
particles vibrating quickly can move away from their fixed positions
a solid material
when heat is added
when particles vibrate quickly the solid becomes a liquid
the temperature at which particles move away from their fixed positions is called the **melting point**
the particles begin to vibrate with even greater speed than before
now the particles vibrating very quickly can break away from each other completely
when particles break away from each other completely, the liquid becomes a gas
a liquid
when heat is added
when particles vibrate very quickly a liquid becomes a gas
the temperature at which particles break away from each other completely is called the **boiling point**

The opposite happens when matter cools. When a gas cools it becomes a liquid. When a liquid cools it becomes a solid.

Evaporation

Evaporation is the word that describes how a liquid slowly turns to a gas at any temperature.

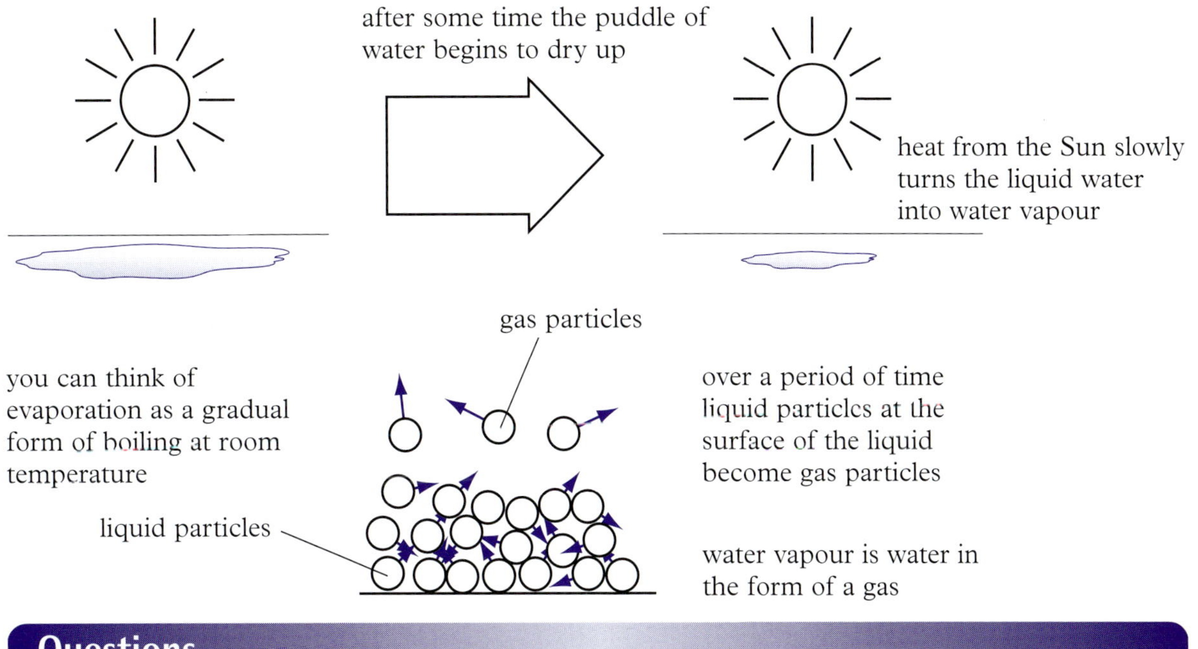
after some time the puddle of water begins to dry up
heat from the Sun slowly turns the liquid water into water vapour
gas particles
you can think of evaporation as a gradual form of boiling at room temperature
over a period of time liquid particles at the surface of the liquid become gas particles
liquid particles
water vapour is water in the form of a gas

Questions

1 The particles in water are attracted to each other.
  a) What happens to the particles when water evaporates?
  b) Why is it easier for the particles to evaporate or escape from the surface of the water?

2 Explain why puddles of water dry up (evaporate more quickly) on a hot day than on a cold day.

3 Why is a windy hot dry day better for drying clothes than a still hot dry day?

4 a) Describe the changes of state which take place when water evaporates.
  b) What causes the changes in state?

Density

If you want to compare how heavy one material is compared with another then you must use **the same volume** of each material. A simple way of doing this would be to compare pieces of the same size and shape.

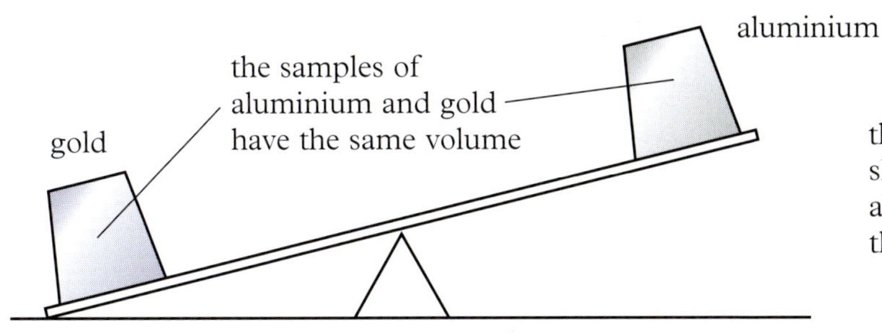
aluminium
the samples of aluminium and gold have the same volume
gold

this see-saw balance shows that this sample of aluminium is lighter than this sample of gold

Density tells us the mass of one cubic centimetre of any material. The units of density are grams per cubic centimetre, shortened to g/cm^3. The diagram below shows the mass of some metals with a volume of 1 cm^3.

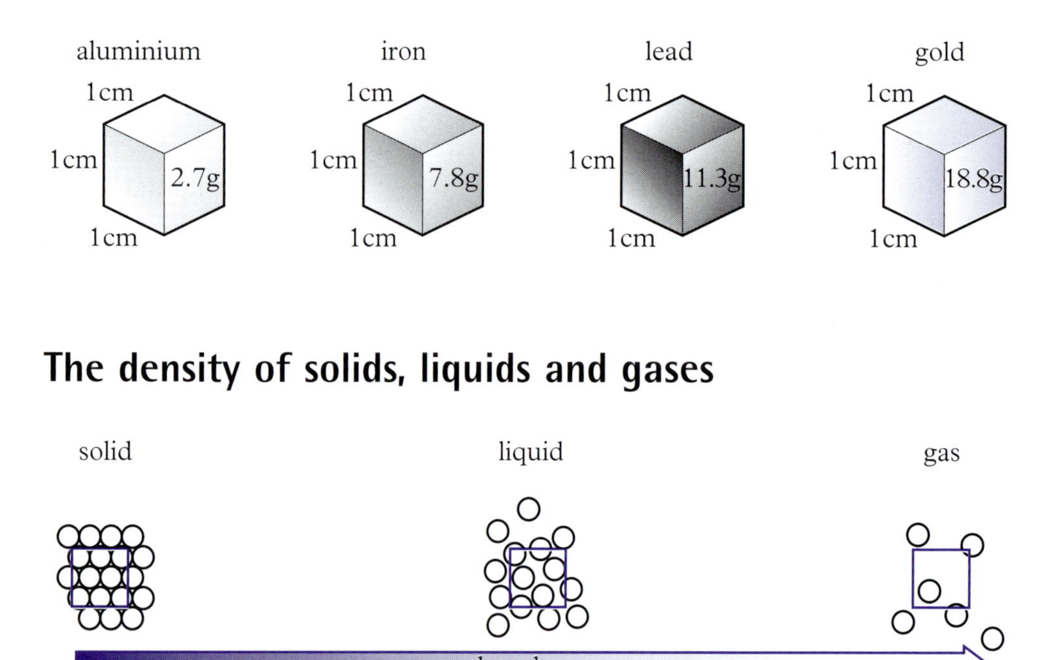
aluminium
1cm
1cm
2.7g
1cm
iron
1cm
1cm
7.8g
1cm
lead
1cm
1cm
11.3g
1cm
gold
1cm
1cm
18.8g
1cm
solid
liquid
gas
less dense

The density of solids, liquids and gases

As a general rule, a material becomes less dense when it melts from a solid to a liquid and very much less dense when it boils to become a gas. You can understand this if you think about the spacing of the particles. Gas particles are much more spread out which means a gas is less dense than a solid and liquid.

Summary

All matter is made of particles. The three states of matter are solid, liquid and gas. The three states are different in the way the particles are arranged and the way the particles move. Solids and liquids have particles that are close together whereas gas particles are far apart.

The particles in a solid vibrate about a fixed position whereas both gases and liquids have particles which are moving at random. When gases, solids and liquids are heated they expand. The particles stay the same size but move further apart. Diffusion is the movement of particles so that they spread out and mix with other particles. Pressure is caused by gas particles hitting the sides of their container. The higher the temperature of a gas, the greater the pressure.

Evaporation is when a liquid slowly turns to a gas as the particles leave the surface of the liquid.

Key words

Boiling point The temperature at which a liquid turns into bubbles of gas.

Diffusion Spreading out of particles.

Evaporation Liquid slowly turning to gas at any temperature.

Expansion Getting bigger.

Freezing point The temperature at which a liquid turns to solid (the same temperature as the melting point for a particular material).

Melting point The temperature at which a solid turns to liquid.

Pressure The force exerted by gases on a particular area of the walls of their containers.

Random Having no fixed position.

Vibrating Moving back and forth about a fixed position.

End of Chapter 1 questions

1 Copy and complete the table below by writing 'yes' or 'no' in the spaces. The first one has been done for you.

	solid	liquid	gas
feels hard to the touch	yes	no	no
can be poured and will fill up a container from the bottom			
keeps its own shape			
always fills all the space in which it is put			
expands in size when heated			
will not stay in an open container			
can be squashed into a smaller volume			
spreads out (diffuses)			
always takes up the same amount of space at a given temperature			

2 Jumping off a 3 metre-high wall onto a concrete floor is likely to hurt you. Jumping into the deep end of a swimming pool will probably not hurt you. Explain why this would happen in terms of the particles in a solid and a liquid.

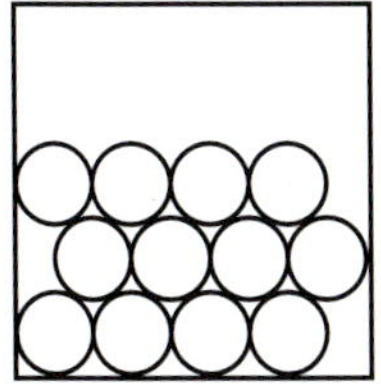

3 The box (right) shows how the particles are arranged in a solid.
a) Draw a box to show how the particles are arranged in a liquid.
b) Draw a box to show how the particles are arranged in a gas.

4 When solids change into liquids (melt) and when liquids change into gases (boil), the material changes in size.

Copy and complete each sentence by choosing a phrase from the following list.

take up less space because the particles are squashed closer together

take up a little more space because the particles move a little further apart

take up much more space because the particles move much further apart

take up the same amount of space

a) When solids melt, they _________________________ .
b) When liquids boil, they _________________________ .

5 How can we change a liquid into a gas?

6 In each case state whether we are talking about a solid, liquid or a gas.
a) In this state of matter the particles are moving very fast and are far apart.
b) In this state of matter the particles are not changing their position.
c) In this state of matter the particles are moving from place to place but are almost touching each other.

7 Which of the diagrams below best represents the following:
a) boiling
b) melting
c) evaporating.

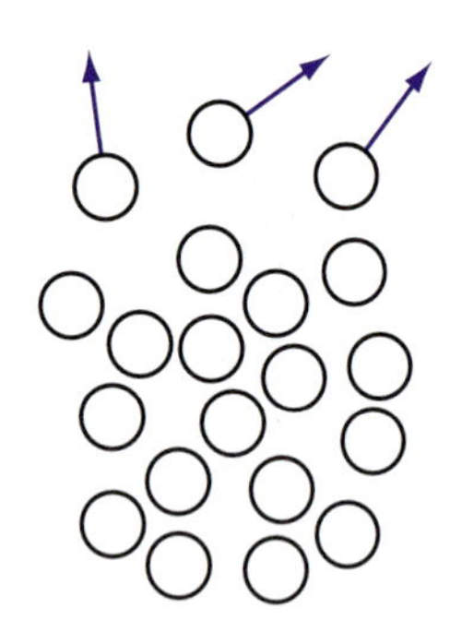

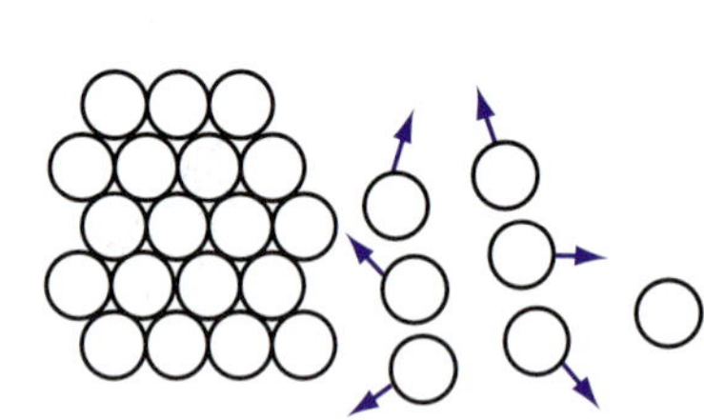

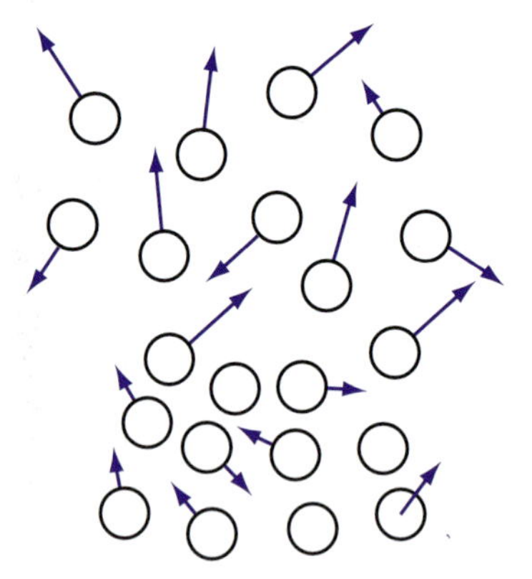

8 When the metal ball and ring are both at room temperature you can just slip the metal ball through the metal ring.

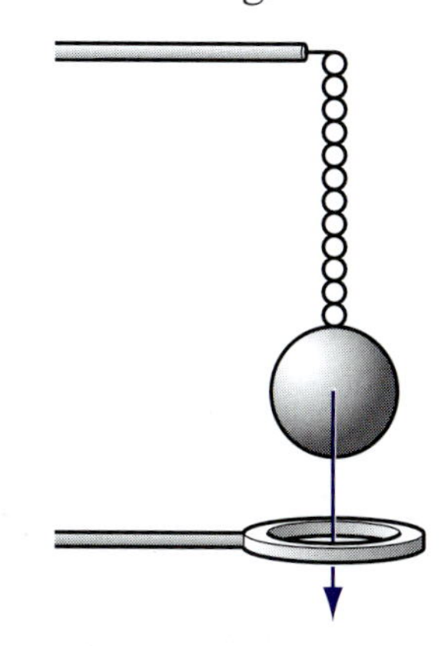

a) What would happen if you heated the metal ball in a Bunsen burner flame and then tried to pass it through the ring?
b) The ball is made of particles. Explain what happened to the particles in the metal ball when it was heated in the Bunsen burner.
c) What would happen if you cooled the ring inside a freezer and then tried to pass the unheated ball through it?
d) Explain your answer to b) by describing what is happening to the particles.

9 Air is made of moving particles. A beach ball pumped up with air feels firm.
a) Explain why the ball feels firm.
b) Say what would happen to the firmness of the ball in each of the following cases. Explain your answer in terms of moving particles.
i) More air is pumped into the ball.
ii) The temperature of the air drops
iii) The temperature of the air goes up.
iv) Some air is let out of the ball.

2 Elements, compounds and mixtures

Elements

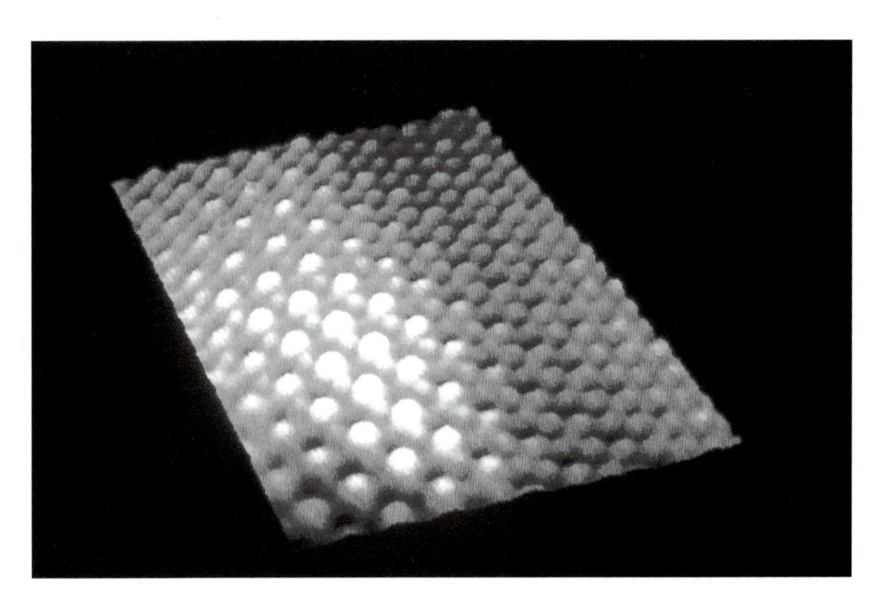

All materials are made of **atoms**. The white circles in the photograph are atoms of a metal called **palladium**. Palladium is an element. An **element** has only one sort of atom.

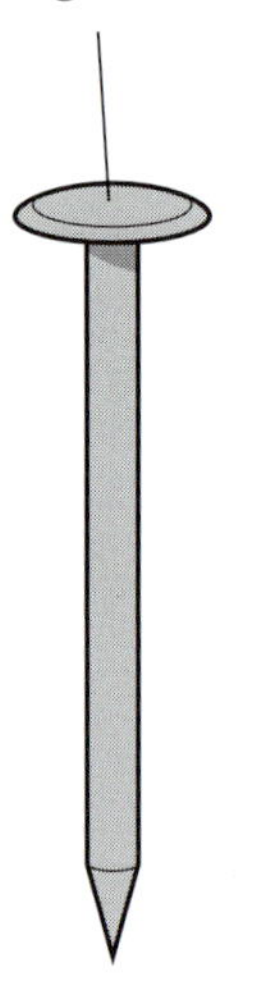
a nail is made of only iron atoms bonded together

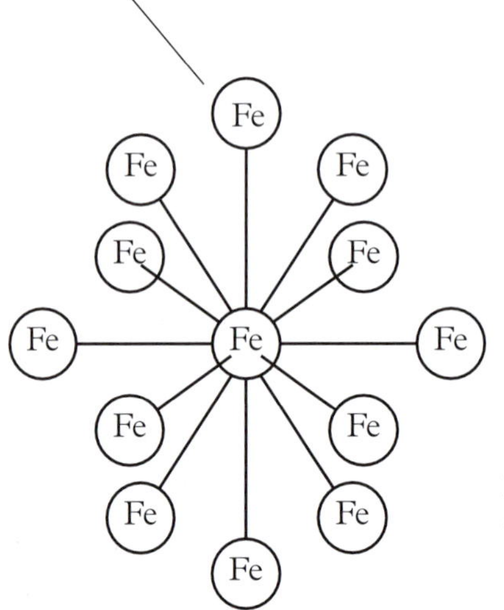
the iron atoms are arranged as shown here
Fe
Fe
Fe
Fe
Fe
Fe
Fe
Fe
Fe
Fe
Fe
Fe
Fe
Fe

Fe is the chemical symbol for an iron atom

a **diamond** is made only of **carbon atoms** bonded together

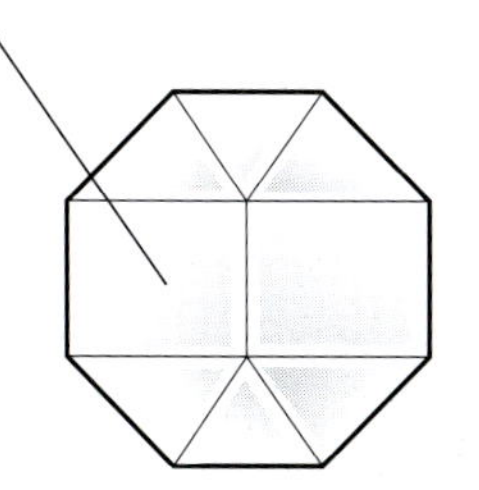

the carbon atoms are arranged as here

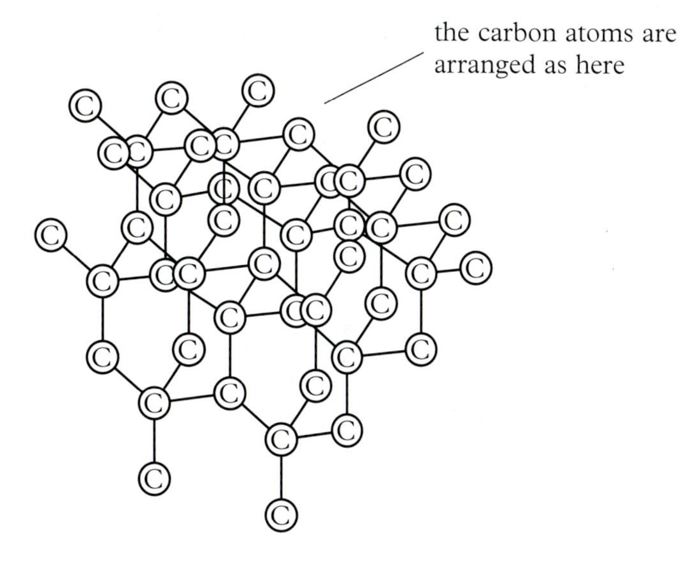

C is the chemical symbol for a carbon atom

Iron and carbon are examples of elements. An element cannot be broken down into anything simpler.

Compounds

Atoms from different elements bond together to make up a compound.

this bottle is made of glass

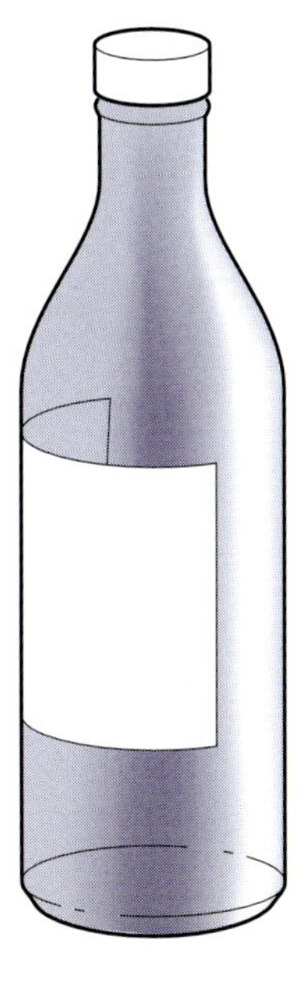

glass is made from calcium, silicon, oxygen and sodium atoms bonded together

the atoms in glass are arranged as here

Ca is the symbol for calcium

Si is the symbol for silicon

O is the symbol for oxygen

Na is the symbol for sodium

these ice cubes are made of water

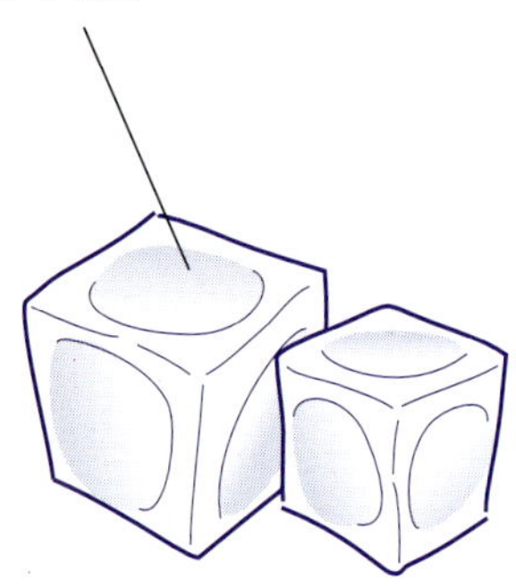

water is made from hydrogen and oxygen atoms bonded together

the atoms in water are arranged as here

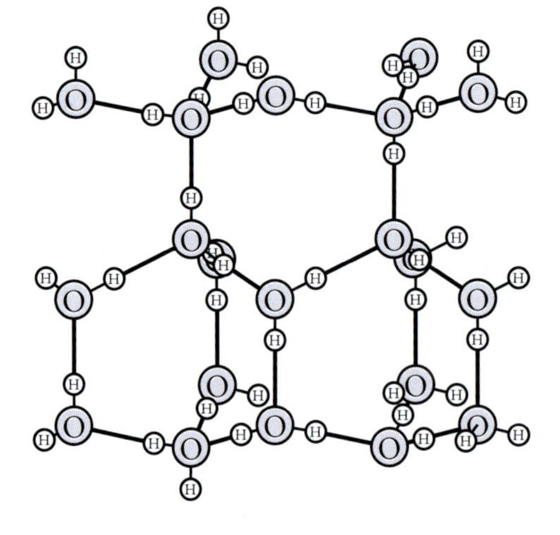
H
O

H is the symbol for hydrogen
O is the symbol for oxygen

Glass and water are both compounds. A compound can be broken down into the elements it is made from. Water for instance can be broken down into hydrogen and oxygen. Hydrogen and oxygen cannot be broken down into anything else because they are elements. Elements and compounds both form single materials.

Remember you cannot see that there are different atoms in glass and in water. The atoms are much too small to see.

Questions

1 What are all materials made from?

2 How many types of atom make up an element?

3 What is the symbol for the element iron?

4 Which element has the symbol C?

5 Copy and complete the table below to show the four elements which make up glass.

element	symbol
calcium	
	O
silicon	
	Na

The elements

There are 109 elements and each of them has a name and a chemical symbol. We have already seen that hydrogen has the symbol H and oxygen the symbol O. The Periodic Table of elements contains all the elements with their symbols.

Here are the first fifty-four elements with their symbols.

H hydrogen																	**He** helium
Li lithium	**Be** beryllium											**B** boron	**C** carbon	**N** nitrogen	**O** oxygen	**F** fluorine	**Ne** neon
Na sodium	**Mg** magne- sium											**Al** alumin- ium	**Si** silicon	**P** phos- phorus	**S** sulphur	**Cl** chlorine	**Ar** argon
K potassium	**Ca** calcium	**Sc** scandium	**Ti** titanium	**V** vanadium	**Cr** chromium	**Mn** manga- nese	**Fe** iron	**Co** cobalt	**Ni** nickel	**Cu** copper	**Zn** zinc	**Ga** gallium	**Ge** germa- nium	**As** arsenic	**Se** selenium	**Br** bromine	**Kr** krypton
Rb rubidium	**Sr** strontium	**Y** yttrium	**Zr** zirconium	**Nb** niobium	**Mo** molyb- denum	**Tc** techne- tium	**Ru** ruthenium	**Rh** rhodium	**Pd** palladium	**Ag** silver	**Cd** cadmium	**In** indium	**Sn** tin	**Sb** antimony	**Te** tellurium	**I** iodine	**Xe** xenon

Chemical formulae

You cannot tell whether a substance is an element or a compound just by looking at it. But, you can sometimes tell from the name. For example, copper oxide is a black powder that looks rather like soot.

Copper oxide has two elements in it, **copper** and **oxygen**. This means it is a compound. The **-ide** on the end means there are only **two elements** in the compound. Copper oxide has one copper atom for every oxygen atom.

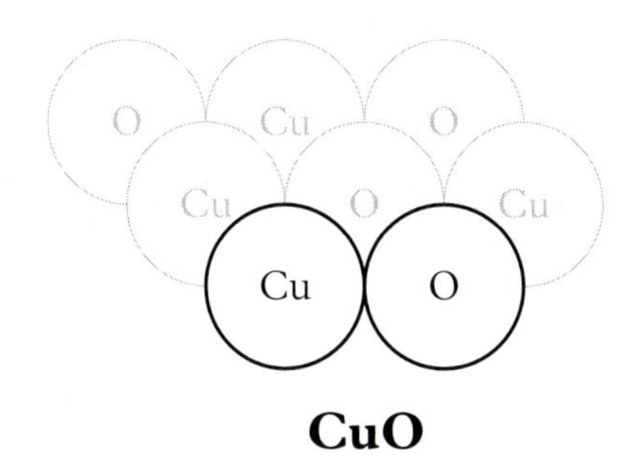
O
Cu
O
Cu
O
Cu
Cu
O
CuO

the chemical symbol CuO is used to show that there is one copper atom for each oxygen atom

CuO is called the **formula** for copper oxide.

What formulae mean

The chemical formula tells us the proportions of atoms which bond together. There is always a fixed ratio of the different elements present in a compound.

Carbon dioxide is made of two oxygen atoms for each carbon atom.

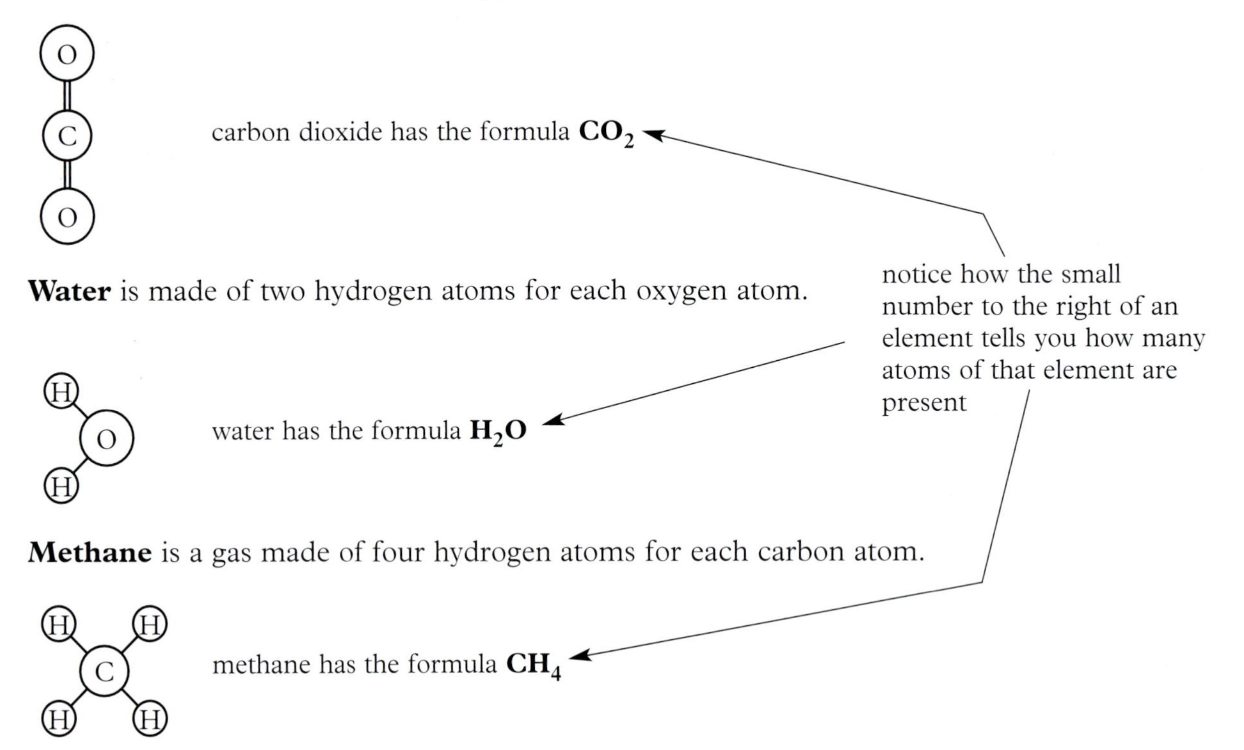
O
C
O
carbon dioxide has the formula CO_2
notice how the small number to the right of an element tells you how many atoms of that element are present
H
O
H
water has the formula H_2O
H
H
C
H
H
methane has the formula CH_4

Water is made of two hydrogen atoms for each oxygen atom.

Methane is a gas made of four hydrogen atoms for each carbon atom.

Sodium chloride is the chemical name for salt that we add to food.

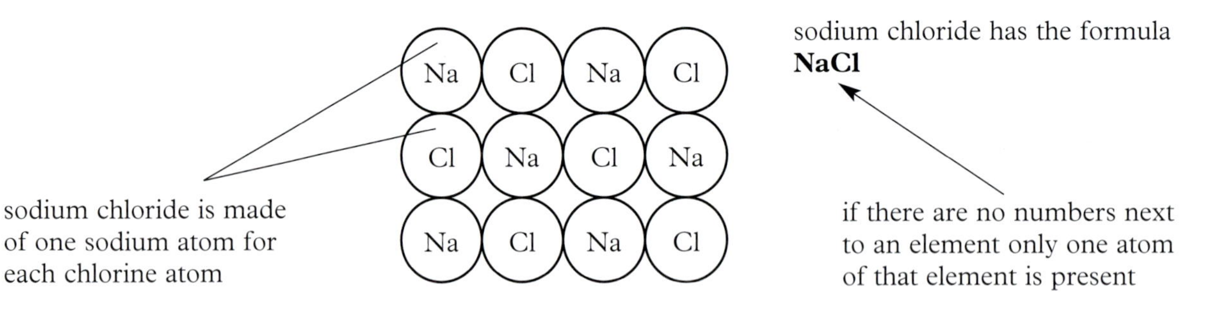
sodium chloride is made of one sodium atom for each chlorine atom
Na Cl Na Cl
Cl Na Cl Na
Na Cl Na Cl
sodium chloride has the formula **NaCl**
if there are no numbers next to an element only one atom of that element is present

Questions

1 Use the Periodic Table to match the symbol to its element

Na	sulphur
O	chlorine
S	oxygen
Pb	lead
Ca	sodium
Cl	calcium

(Hint: In the past **pl**um**b**ers used lead for pipes).

2 Work out which substances in the list are elements and which are compounds. Copy and complete the table with the names of the elements and compounds.

oxygen	**water**	**sodium**
carbon dioxide	**methane**	**hydrogen**
copper oxide	**calcium**	**carbon**

element	compound

Elements which are gases

Elements may be solids, liquids or gases. Many of the elements that are gases exist as pairs of atoms:

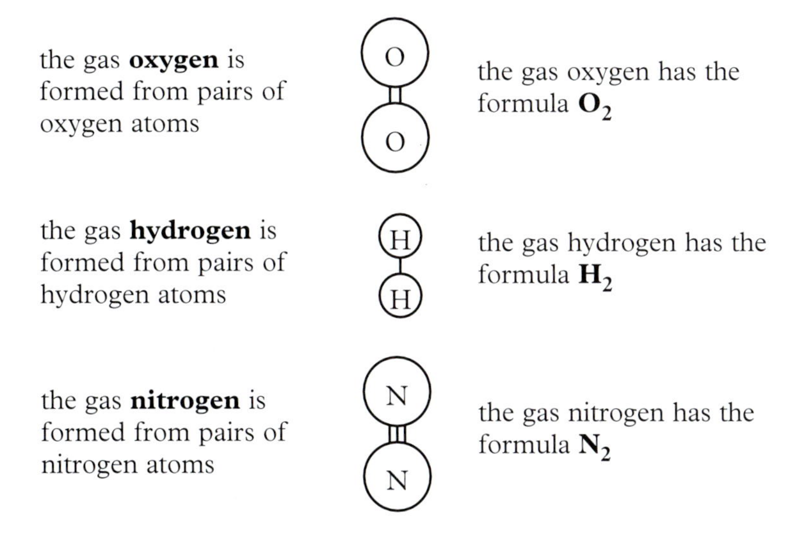

the gas **oxygen** is formed from pairs of oxygen atoms
O
O
the gas oxygen has the formula $\mathbf{O_2}$
the gas **hydrogen** is formed from pairs of hydrogen atoms
H
H
the gas hydrogen has the formula $\mathbf{H_2}$
the gas **nitrogen** is formed from pairs of nitrogen atoms
N
N
the gas nitrogen has the formula $\mathbf{N_2}$

Here are some formulae to learn by heart:

formula	name
H_2	hydrogen
N_2	nitrogen
O_2	oxygen

formula	name
H_2O	water
HCl	hydrochloric acid
NaCl	sodium chloride
CO_2	carbon dioxide

Compounds and mixtures

Mixtures and compounds both contain at least two different elements.

Compounds

A compound is a single substance, either a gas, liquid or a solid. It is made up from two or more atoms strongly bonded together.

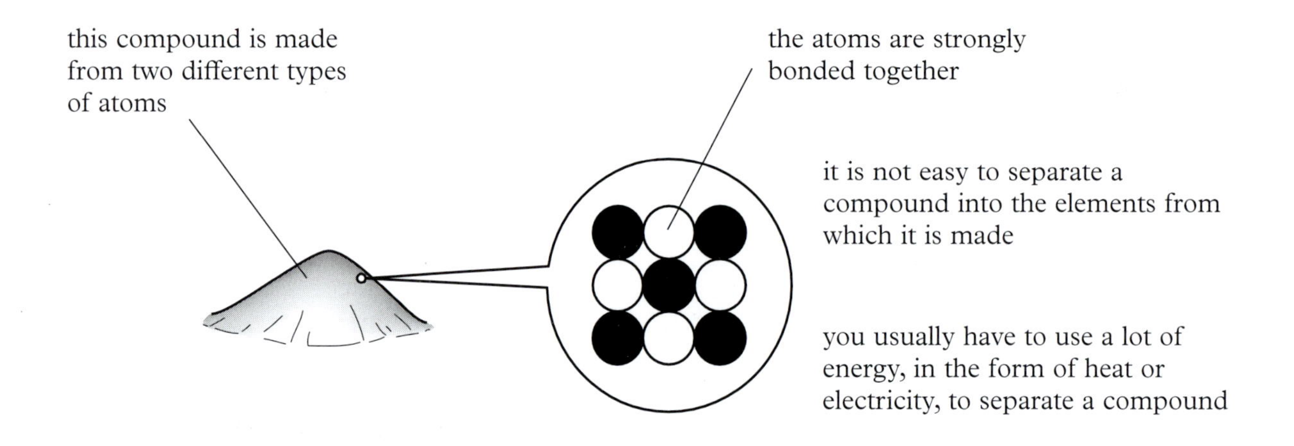
this compound is made from two different types of atoms
the atoms are strongly bonded together
it is not easy to separate a compound into the elements from which it is made
you usually have to use a lot of energy, in the form of heat or electricity, to separate a compound

Mixtures

A mixture consists of two or more separate substances that have been added to each other. The atoms in one substance are not bonded to the atoms in the other substance.

Mixtures around us

There are all sorts of mixtures. Here are some examples:

air is mainly made of nitrogen and oxygen

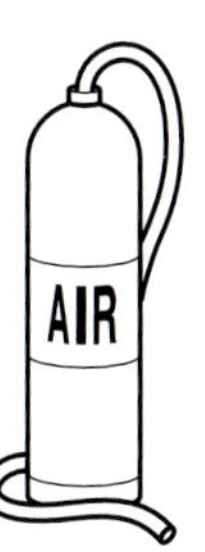
AIR

in air the gas nitrogen is mixed with the gas oxygen

sparkling water is made from carbon dioxide and water

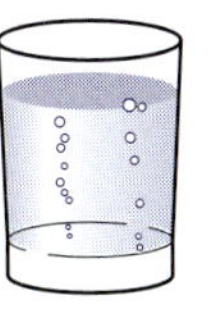

in sparkling water the gas carbon dioxide is mixed with liquid water

a pumice stone has air pockets in holes in the rock

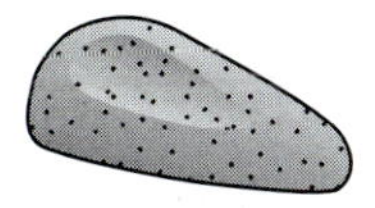

a pumice stone is the gas air mixed with a solid rock

wine is made from alcohol and water

wine is the liquid alcohol mixed with liquid water

sea water is made from water mixed with salt

sea water is liquid water mixed with salt which is a solid

the solid is dissolved in the water, so you can't actually see it

soil is made from clay and sand

soil is a solid mixed with a solid

Separating mixtures

Because the substances are not bonded together, **mixtures can usually be separated easily**.

To separate the different substances in a mixture you need to find a difference in the way the substances behave. Substances behave differently because of a difference in their properties.

Physical separation

To separate iron filings from talcum powder you could **use a magnet**.

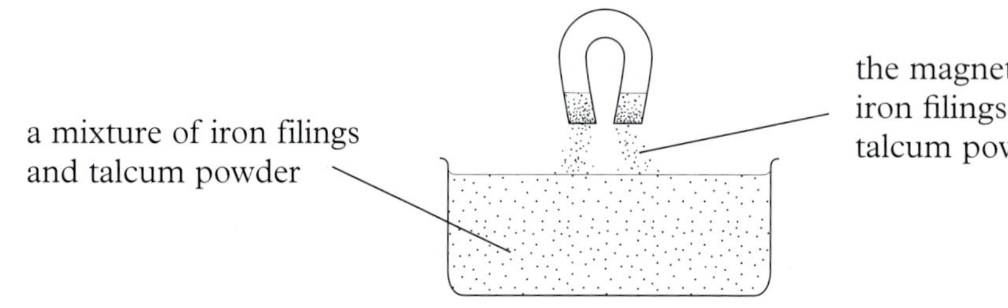
a mixture of iron filings and talcum powder
the magnet attracts the iron filings but not the talcum powder

Filtration

To separate sand and water, you could use **filtration**.

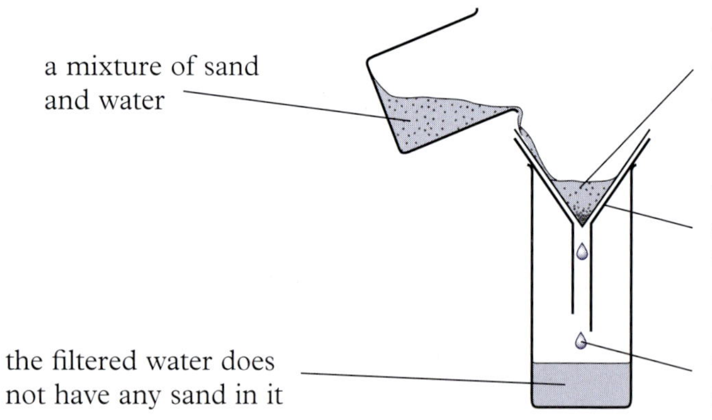
a mixture of sand and water
the mixture of sand and water is poured through some filter paper
the filter paper only allows water to pass through it
the filtered water does not have any sand in it
the water is separated from the sand

Distillation

To separate the *water* from a salt and water mixture you could use **distillation**.

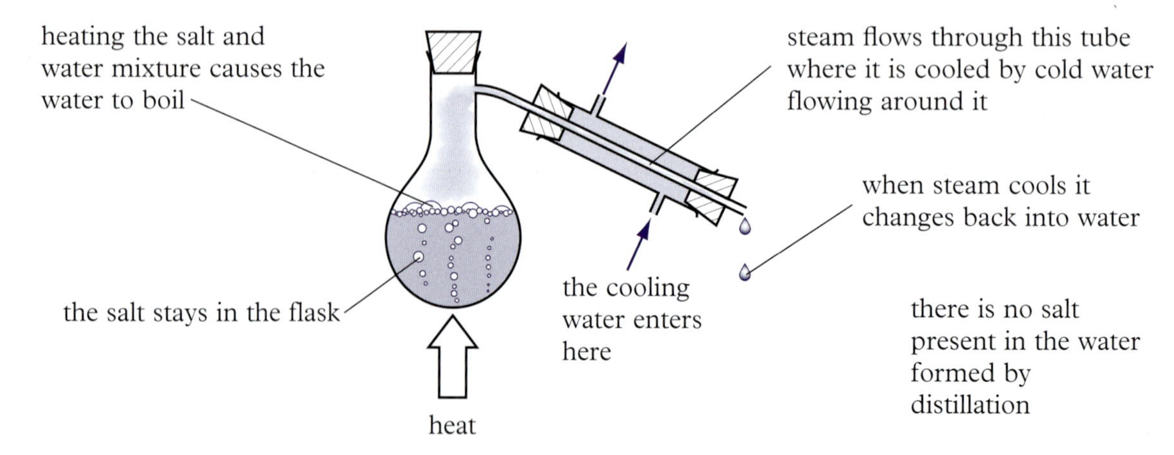
heating the salt and water mixture causes the water to boil
steam flows through this tube where it is cooled by cold water flowing around it
when steam cools it changes back into water
the salt stays in the flask
the cooling water enters here
there is no salt present in the water formed by distillation
heat

Evaporation

To separate the *salt* from salt and water, you could use **evaporation**.

heating the salt and water mixture causes the water to boil and change to steam

when the water changes to steam it escapes from the dish

eventually all the water boils away and only salt remains in the dish

heat

Chromatography

Ink is made up of a mixture of different coloured dyes. The dyes can be separated using a process called **chromatography**.

the ink spot is a mixture of different dyes

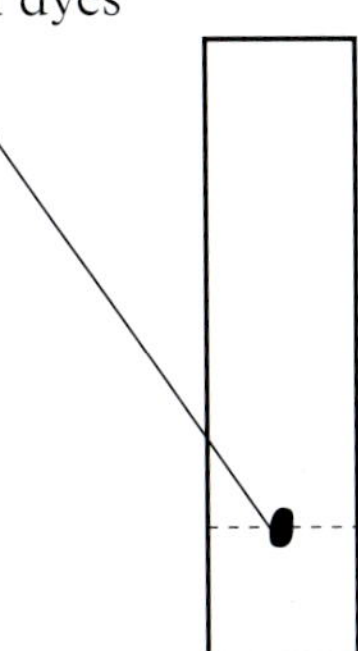

the bottom of the filter paper is dipped into water

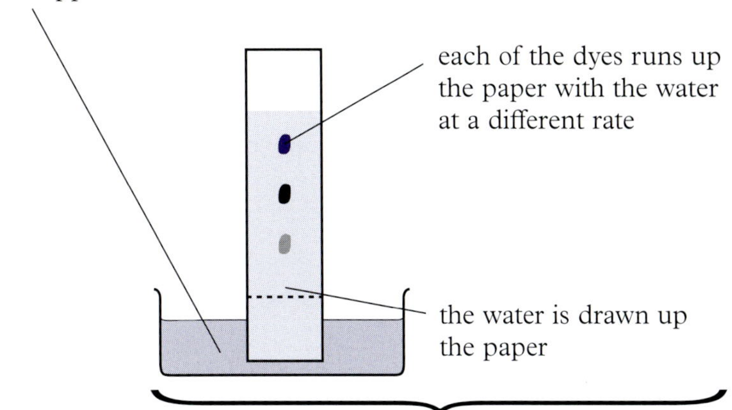
each of the dyes runs up the paper with the water at a different rate
the water is drawn up the paper

in this way the dyes separate from the ink

Questions

1. Explain how you could separate a mixture of metal paper clips from plastic ones.
2. Give two ways of separating salt from sea water.
3. How could you show that blue ink dye is a mixture of different coloured dyes?
4. Describe how clean water could be obtained from muddy water.

The main differences between compounds and mixtures

Compounds	Mixtures
A compound is a single substance.	A mixture contains more than one sort of substance.
It is not easy to separate a compound into the elements from which it is made.	The substances in a mixture can usually be separated easily.
There is always a fixed ratio of the elements present in a compound.	The amount of each substance in a mixture is not fixed, we can have any amount of each.
A compound is a single substance with its own set of properties.	A mixture will have the properties of the substances that it is made from. For example a mixture of salt and sugar tastes both sweet and salty.

Summary

Elements contain only one sort of atom. Compounds have more than one sort of atom bonded together.

The Periodic Table contains all the elements with their chemical symbols.

The formula of a substance tells us which atoms are present. It also tells us the simplest ratios of each element present. In a compound there is always a fixed ratio of elements. The elements in a compound are not easy to separate.

A mixture can have any amount of substances present. A mixture can be separated because of the difference in properties of the substances from which it is made.

Key words

compound A gas, liquid or solid containing two or more elements *bonded* together.

element A gas, liquid or solid consisting of just one sort of atom.

mixture A gas, liquid or solid containing more than one different compound or element *not bonded* together.

End of Chapter 2 questions

1 a) Copy the diagrams below and label each one from the following list:

oxygen nitrogen nitrogen dioxide mixture of nitrogen and oxygen

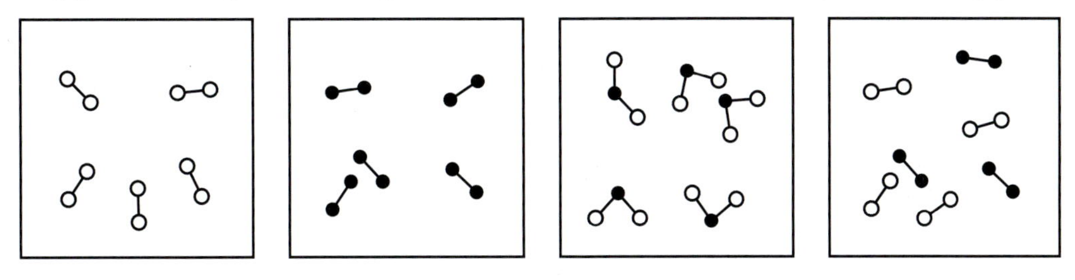

b) Write below each diagram whether the substance is a single element, a compound or a mixture.

c) Give three differences between a mixture and a compound.

2 Use the Periodic Table to help you complete the table below.

substance	formula	elements present: name of element	elements present: number of atoms
water	H_2O		
ammonia	NH_3		
aluminium oxide	Al_2O_3		
copper sulphate	$CuSO_4$		

3

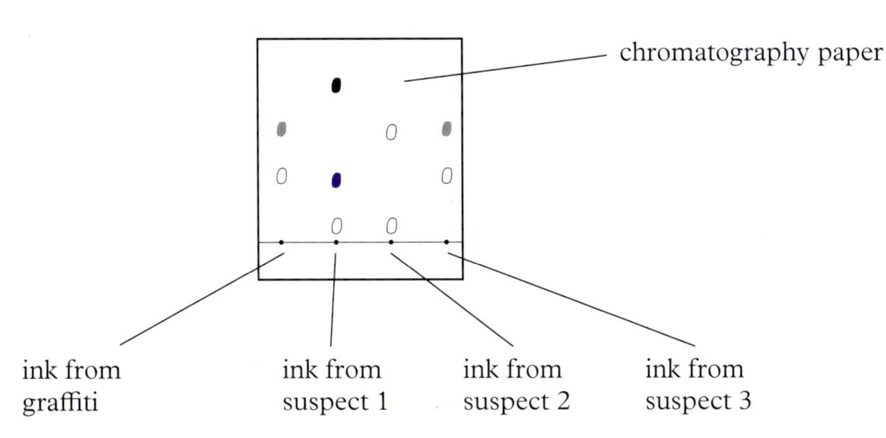

chromatography paper
ink from graffiti
ink from suspect 1
ink from suspect 2
ink from suspect 3

The diagram shows the result of an experiment to compare the black ink used to draw graffiti with the inks found in the pens belonging to three suspects.

a) What is the name of the technique used in this experiment?

b) Explain briefly how you would carry out the experiment to get the results shown in the diagram.

c) Which suspect could have written the graffiti?

d) How many different coloured dyes are there in the ink from the first suspect's pen?

e) All the inks contain more than one dye. Are the inks mixtures or compounds? Explain your answer.

3 Atomic structure

The idea that everything was made of atoms came from a Greek philosopher called Democritus who lived about 2000 years ago. The story is that he imagined cutting up a single grain of sand over and over again until he reached a particle that was so small that it couldn't be divided. He called this the **atom**.

At the time his ideas were not accepted. Then, about 1700 years later scientists decided that matter was made up of **elements**. At that time it was thought that substances could not be broken down into anything else. Later still scientists agreed that this meant elements were made of atoms.

The atoms of elements

Every element is made up of atoms that are different from the atoms of any other element.

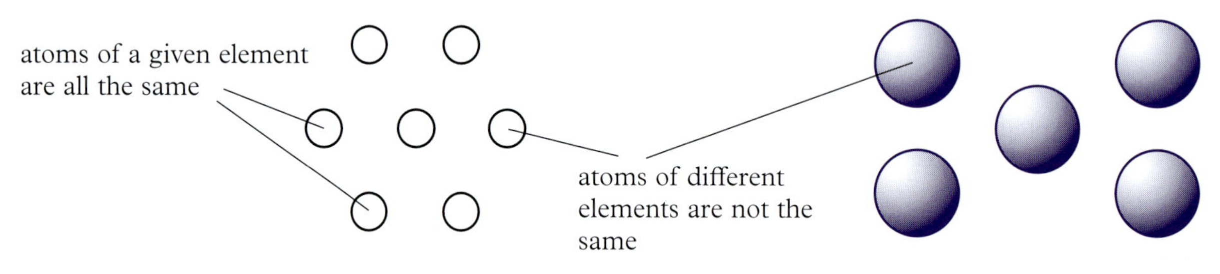
atoms of a given element are all the same
atoms of different elements are not the same

Inside the atom

The atoms that make up all materials are so small that we cannot see one even with the most powerful microscope. Yet atoms themselves are made up of even smaller particles.

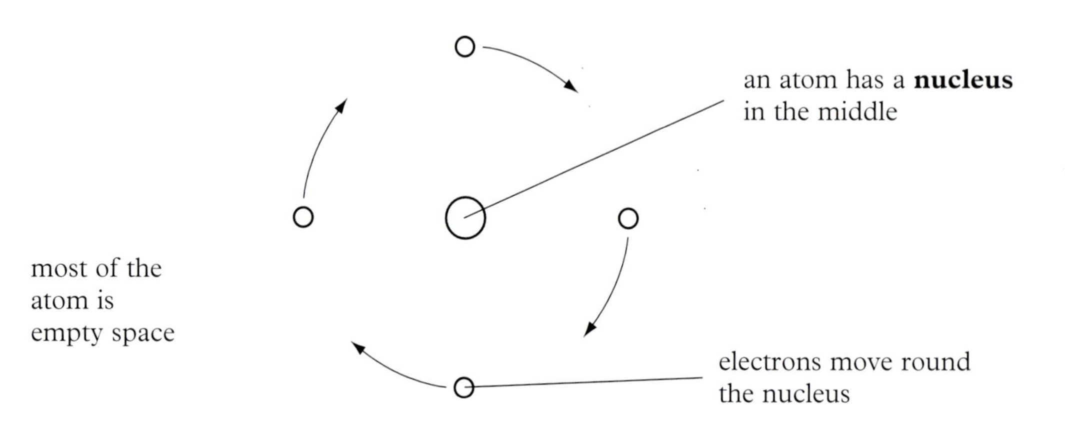
an atom has a **nucleus** in the middle
most of the atom is empty space
electrons move round the nucleus

If the nucleus were the size of a football in the centre circle of a football pitch, the electrons would be at the edge of the football stadium.

The nucleus

nearly all the mass of an atom is in the nucleus

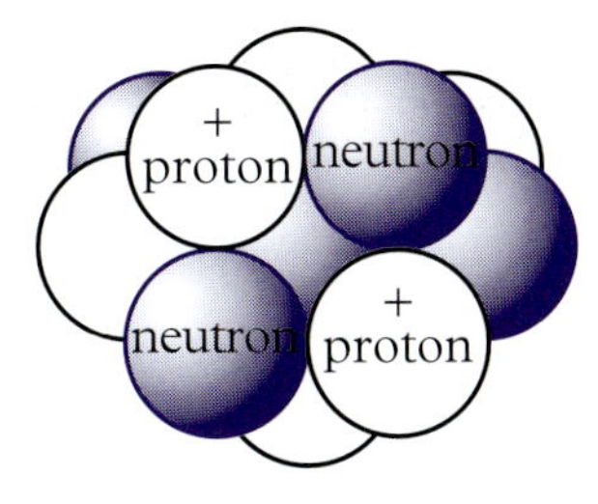
+
proton
neutron
+
proton
neutron

the main particles that make up the nucleus are **protons** and **neutrons**

Electric charge

There are two sorts of electric charge; positive (+) and negative (−). Like charges repel each other (**+**)←→(**+**) or (**−**)←→(**−**). Opposite charges attract (**+**)→←(**−**). If an object has the same amount of positive and negative charges, it is neutral.

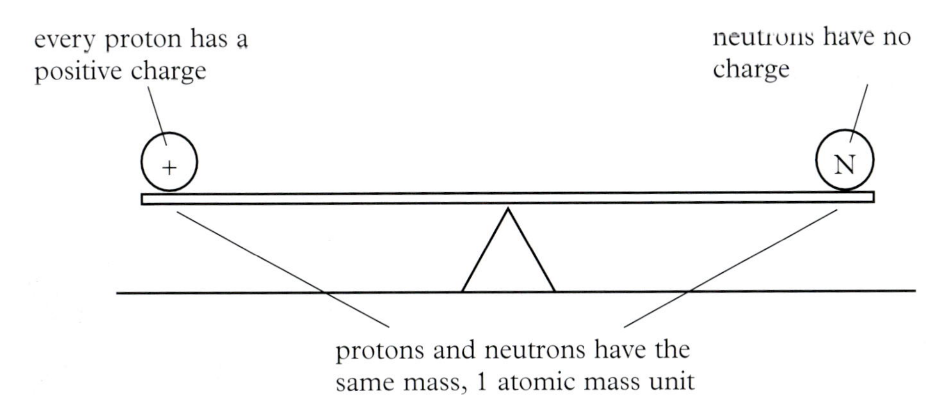
every proton has a positive charge
neutrons have no charge
+
N
protons and neutrons have the same mass, 1 atomic mass unit

Electrons

Electrons are very light and negatively charged.

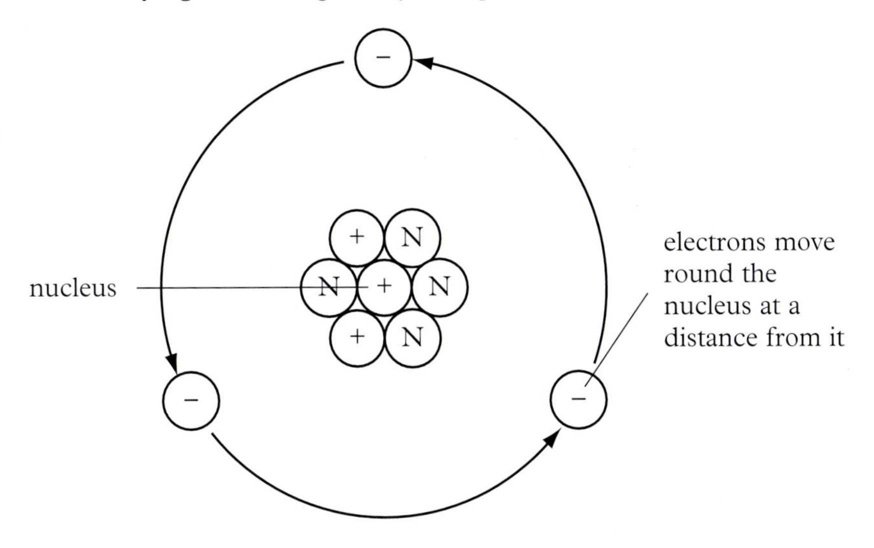
nucleus
electrons move round the nucleus at a distance from it
+
N
+
N
+
N
N
−
−
−

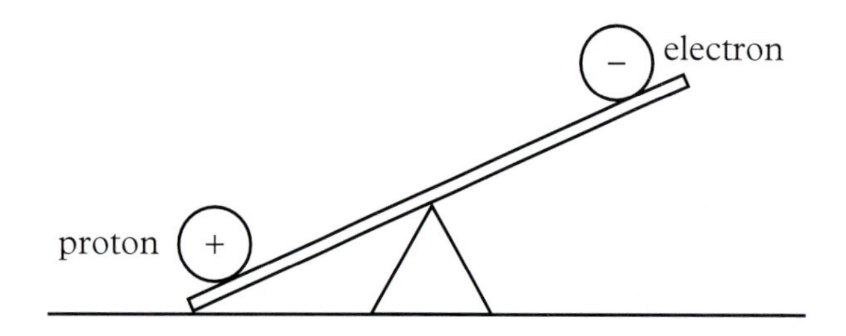
electron
proton

electrons have almost no mass

every electron has a negative charge (−1) which is equal and opposite to the charge on a proton (+1)

Atoms are **neutral**, which means they have no overall positive or negative charge. This is because the number of positive protons is equal to the number of negative electrons. For example, in the helium atom:

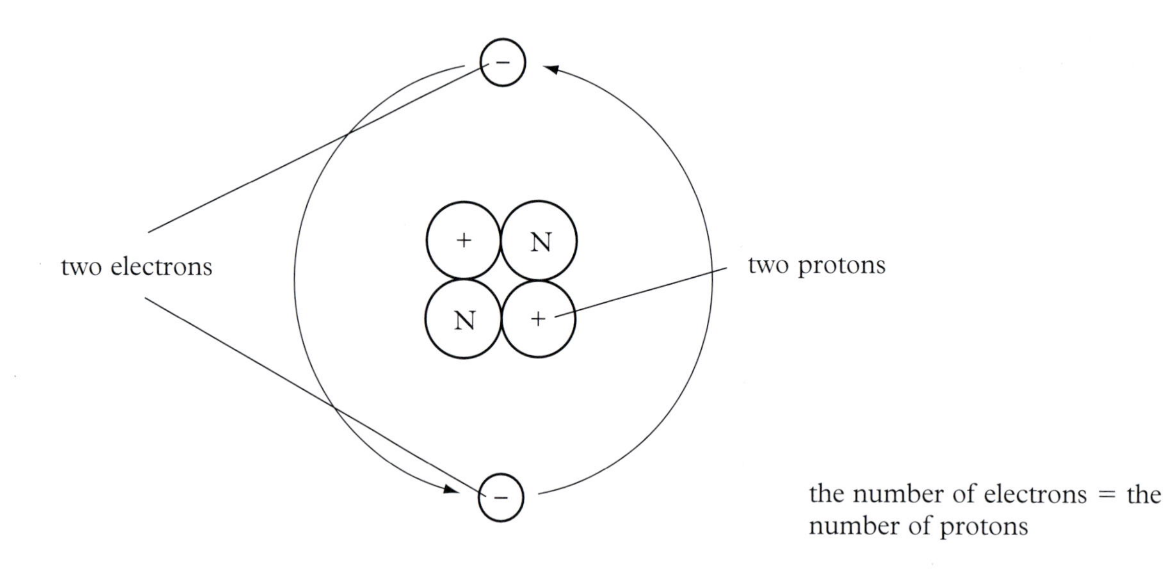
two electrons
two protons
N
N

the number of electrons = the number of protons

Questions

1 Copy and complete the table below.

	Proton	Neutron	Electron
Mass	1 atomic mass unit		Almost nothing
Charge	+1	neutral	
Position in the atom		in the nucleus	moving round the nucleus

2 An atom is neutral.
How do you know that the number of electrons is equal to the number of protons in an atom?

The atomic number

Every element has its own atomic number. This is the number of protons in the nucleus.

All the atoms of a given element have the same number of protons in the nucleus.

The number of protons is called the atomic number of the element.

This number is different for every element.

The atomic number of hydrogen
The simplest atom of all, and the lightest, is hydrogen, atomic number 1.

hydrogen has no
neutrons in the nucleus

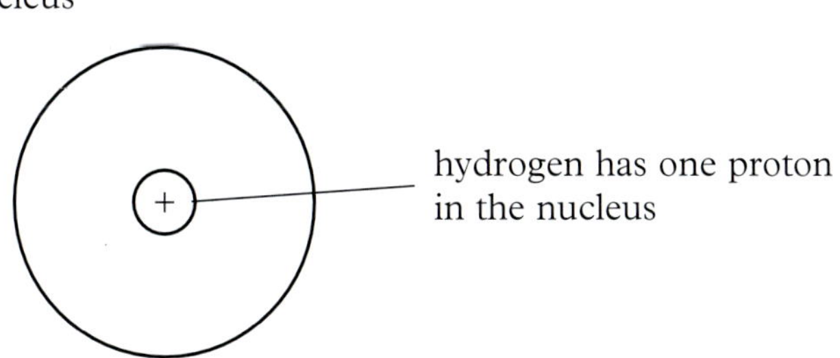
+
hydrogen has one proton
in the nucleus

The atomic number of helium
The second simplest atom is helium, atomic number 2.

helium has two neutrons
in the nucleus

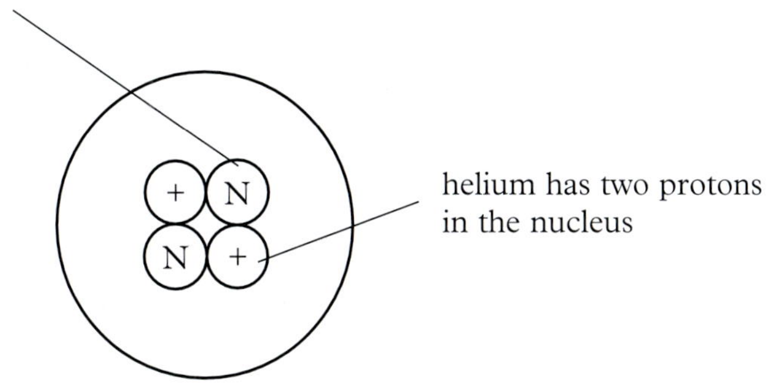
+
N
N
+
helium has two protons
in the nucleus

The atomic number of lithium

The third simplest atom is lithium, atomic number 3.

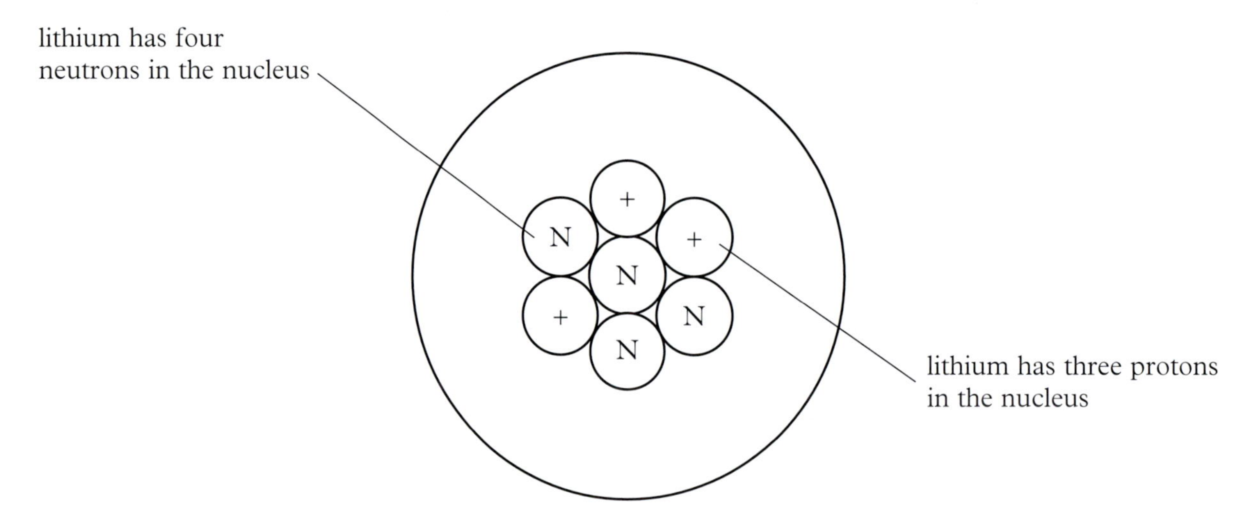
lithium has four neutrons in the nucleus
lithium has three protons in the nucleus
N
+
N
+
N
+
N
N

Lithium is a very simple atom. The largest atom that we know has 109 protons in every atom and is called meitnerium.

The mass number

If we add together the number of protons and the number of neutrons we can find the mass of one atom of the element. This number is called the **mass number**.

mass number = number of protons + number of neutrons

For example, an atom of beryllium has a mass number of 9.

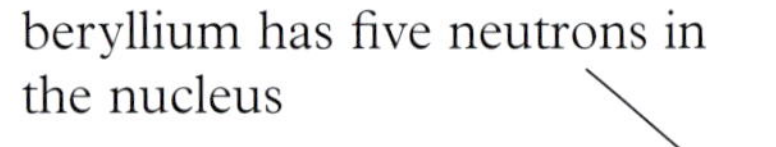
beryllium has five neutrons in the nucleus

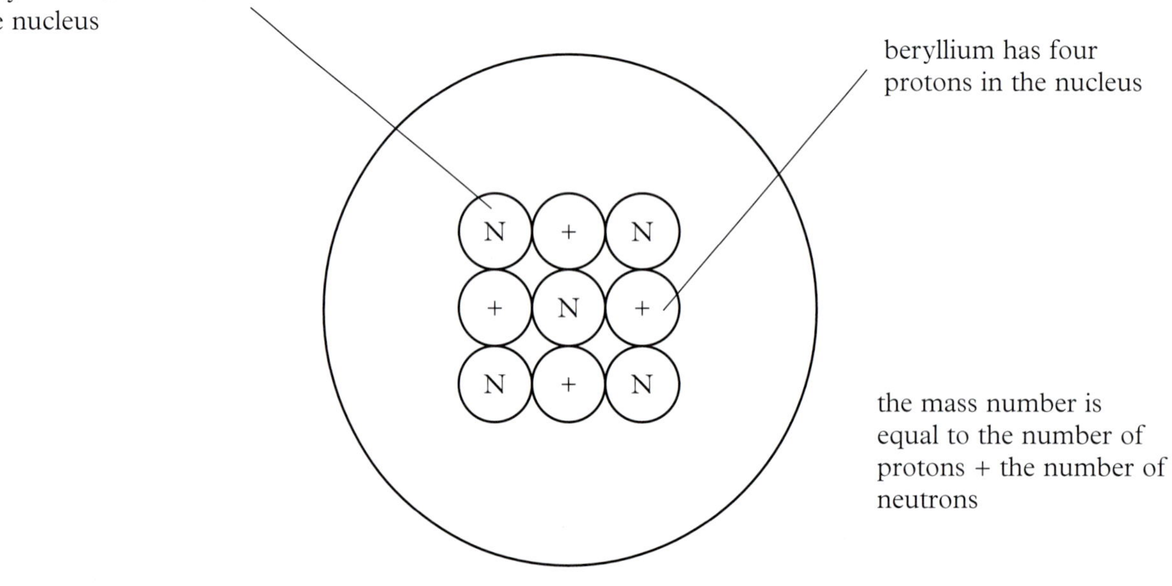
beryllium has four protons in the nucleus
N
+
N
+
N
+
N
+
N
the mass number is equal to the number of protons + the number of neutrons

the mass number of beryllium is 9

Representing mass number and atomic number

Atoms of an element are often written in shorthand to show the mass number and the atomic number.

For example, for **lithium** which has the chemical symbol **Li**:

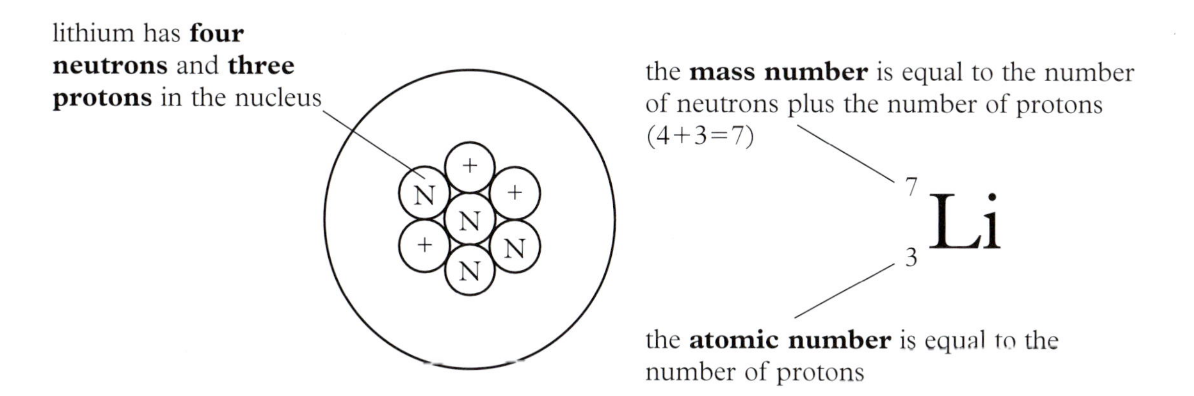
lithium has **four neutrons** and **three protons** in the nucleus
N
+
N
+
+
N
N
the **mass number** is equal to the number of neutrons plus the number of protons (4+3=7)
7
Li
3
the **atomic number** is equal to the number of protons

Helium, which has the chemical symbol He, has **two neutrons** and **two protons** in the nucleus

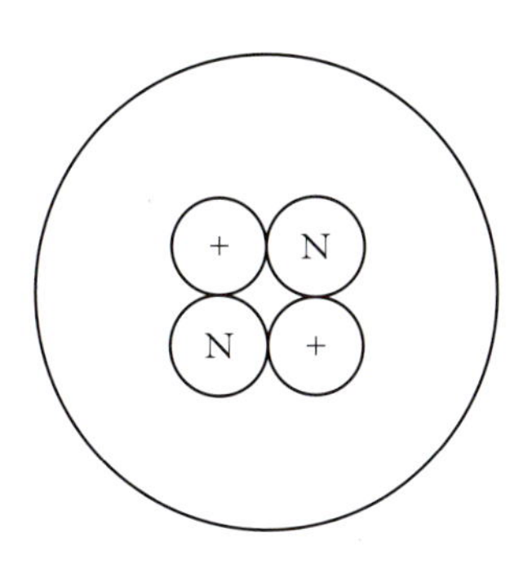
+
N
N
+

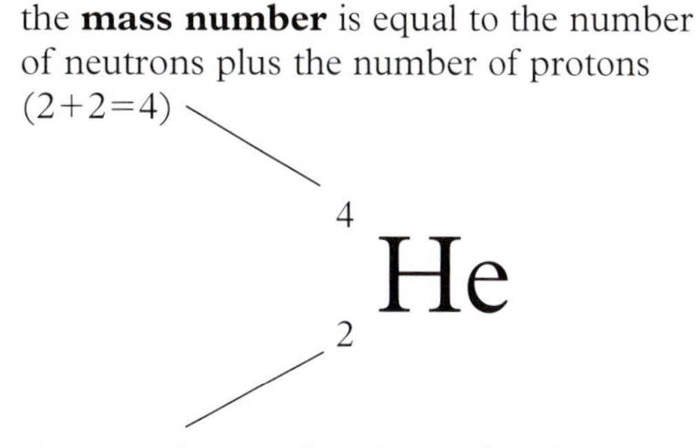
the **mass number** is equal to the number of neutrons plus the number of protons (2+2=4)
4
He
2
the **atomic number** is equal to the number of protons

The mass number is equal to the number of neutrons and protons. The atomic number is equal to the number of protons.

If we take the atomic number away from the mass number we are left with number of neutrons present in that atom.

For the two elements we have just seen:

$^{7}_{3}Li$ the number of neutrons = mass number − atomic number
the number of neutrons = 7 − 3
the number of neutrons = 4

If we look at the diagram of lithium we can see that this is so.

$^{4}_{2}He$ the number of neutrons = mass number − atomic number
the number of neutrons = 4 − 2
the number of neutrons = 2

If we look back at the diagram of helium we can see that this is so.

The first 20 elements

Here are the first 20 elements in order of atomic number.

1 H hydrogen 1							4 He helium 2
7 Li lithium 3	9 Be beryllium 4	11 B boron 5	12 C carbon 6	14 N nitrogen 7	16 O oxygen 8	19 F flourine 9	20 Ne neon 10
23 Na sodium 11	24 Mg magnesium 12	27 Al aluminium 13	28 Si silicon 14	31 P phosphorus 15	32 S sulphur 16	35 Cl chlorine 17	40 Ar argon 18
39 K potassium 19	40 Ca calcium 20						

Questions

1 Use the diagrams of the atoms on pages 29–30 to work out the mass number for:
   a) hydrogen
   b) helium
   c) lithium.

2 Make a table like the one below and fill in the columns for the first 20 elements. The first two have been done for you.

Element	Number of protons (atomic number)	Number of protons + number of neutrons (mass number)	Number of neutrons (mass number – atomic number)
$^{1}_{1}H$	1	1	0
$^{4}_{2}He$	2	4	2

Isotopes

Not all the atoms of an element are **exactly** the same. They all have the same number of protons (the atomic number) otherwise they would not be the same element. Some elements have atoms with different numbers of neutrons in their nuclei. This means that they have different mass numbers. These different atoms of the same element are called **isotopes**.

Carbon has several different isotopes. The most common form of carbon is written as ${}^{12}_{6}\text{C}$.

the most common form of carbon has **6 neutrons** and 6 protons

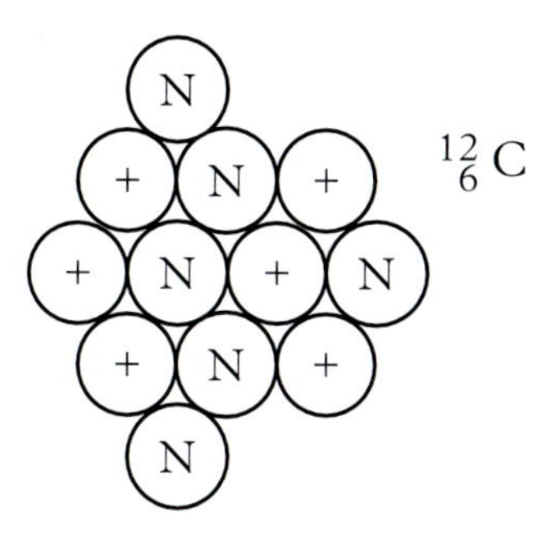

${}^{12}_{6}\text{C}$

a form of carbon exists with 7 **neutrons**

another form of carbon has **8 neutrons**

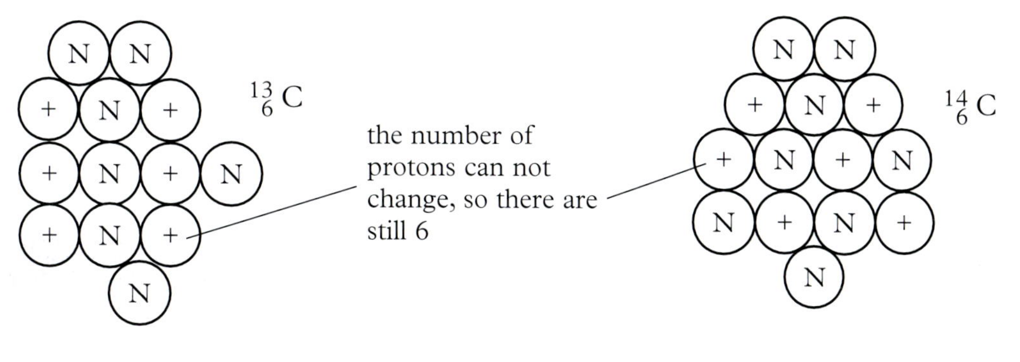

${}^{13}_{6}\text{C}$
the number of protons can not change, so there are still 6
${}^{14}_{6}\text{C}$

Relative atomic mass

The **relative atomic mass** is another way of comparing the masses of atoms. It takes into account all the isotopes of that element, so it may be slightly different from the mass number. It is an average of all the mass numbers of all the isotopes of an element. It is often shortened to $\mathbf{A_r}$. For example, chlorine has a relative atomic mass of 35.5 because chlorine is a mixture of isotopes. For every three atoms with a mass number of 35, there is one with a mass number of 37. For many elements, the mass number of the most common isotope and the relative atomic mass are almost the same – especially if we are working to the nearest whole number.

The way in which the electrons go round the nucleus

Electrons are tiny negatively-charged particles that go round (or **orbit**) the nucleus. There are always the same number of electrons as protons, so an element with an atomic number of 10 has 10 protons and also 10 electrons. There is a pattern in where the electrons are found.

Every atom is a little like our solar system. In the middle, like the Sun, is the nucleus. Circling round the nucleus, like the planets round the Sun, are the electrons. Electrons circle round the nucleus at certain distances away from the nucleus. The positions the electrons can occupy are called **shells**.

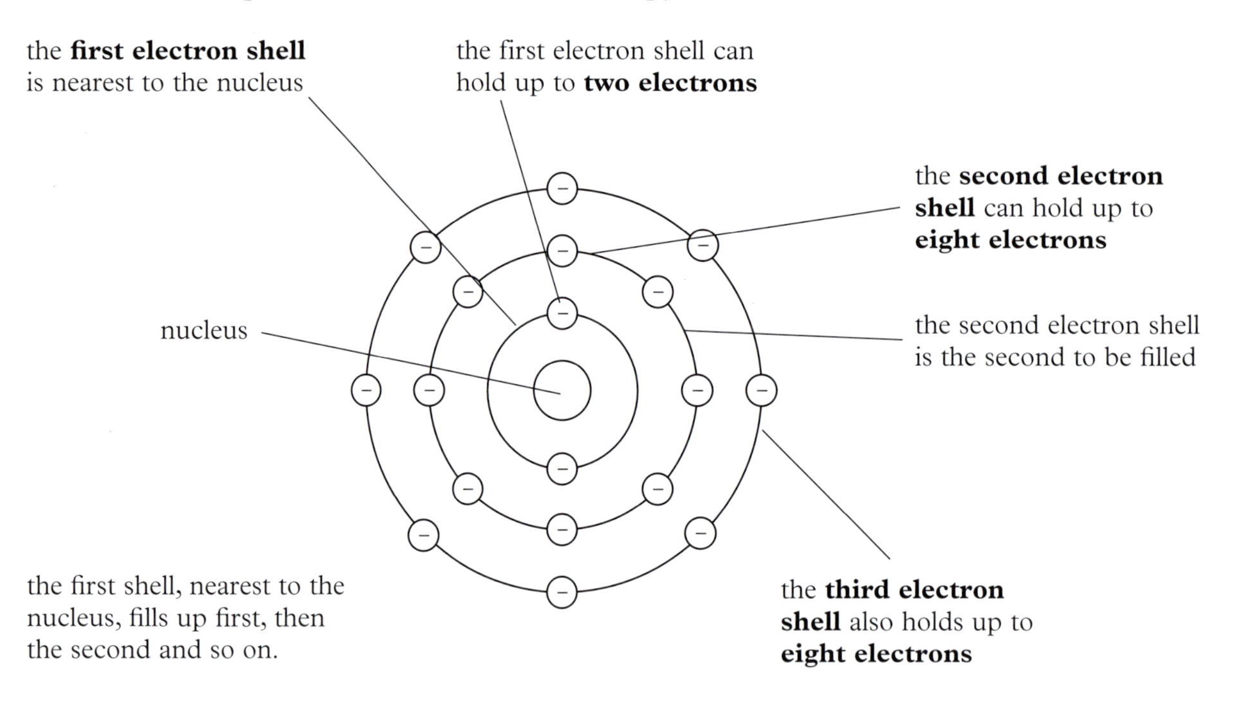
the **first electron shell** is nearest to the nucleus
the first electron shell can hold up to **two electrons**
the **second electron shell** can hold up to **eight electrons**
nucleus
the second electron shell is the second to be filled
the first shell, nearest to the nucleus, fills up first, then the second and so on.
the **third electron shell** also holds up to **eight electrons**

The number of electrons

Different elements have different numbers of electrons. The positions of electrons differ between different elements.

The table of elements on page 32 lists lithium as $^{7}_{3}\mathbf{Li}$. From this we know its atomic number is 3. This means we know it has 3 protons and also 3 electrons.

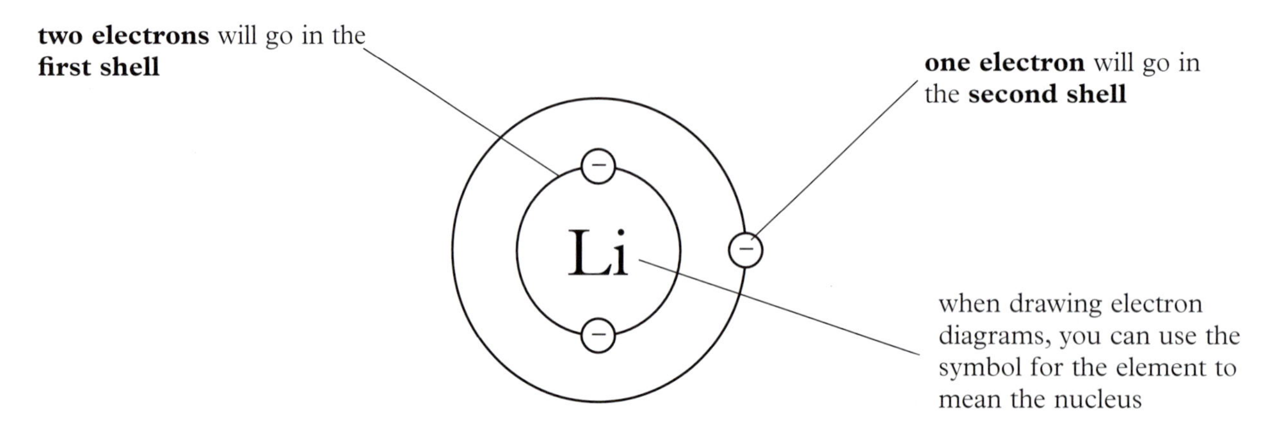
two electrons will go in the **first shell**
one electron will go in the **second shell**
Li
when drawing electron diagrams, you can use the symbol for the element to mean the nucleus

There are two electrons in the first shell and one electron in the second shell. This electron arrangement may be shortened to **2,1**.

Boron

The table of elements on page 32 lists boron as $^{11}_{5}\mathbf{B}$. From this we know its atomic number is **5**. This means we know it has 5 protons and also 5 electrons.

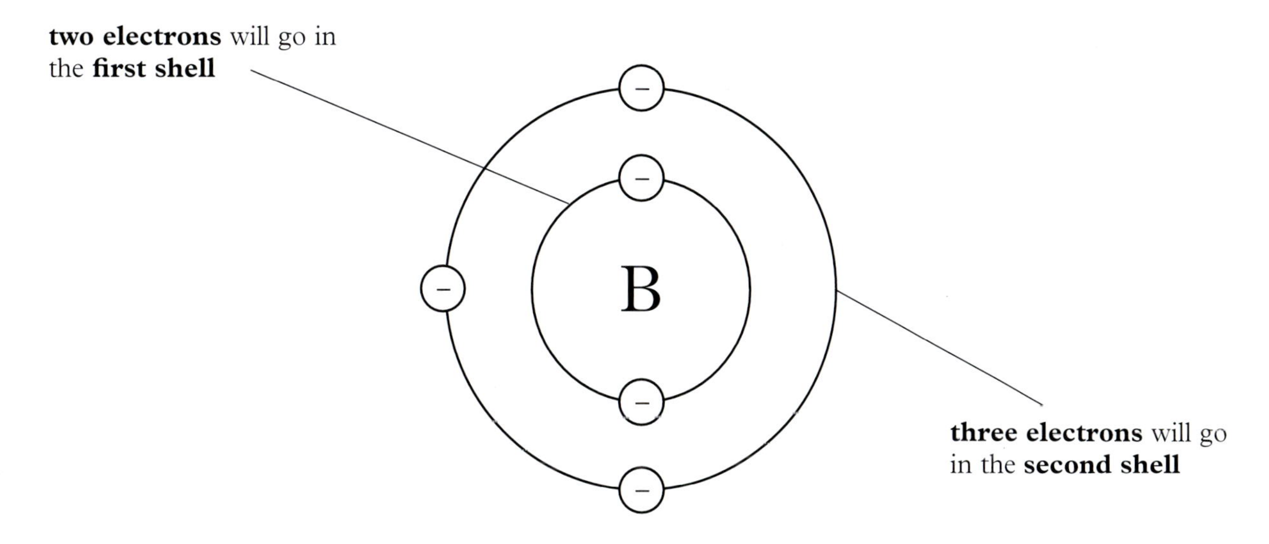
two electrons will go in the **first shell**
B
three electrons will go in the **second shell**

Boron has two electrons in the first shell and three electrons in the second shell. You can write this arrangement as **2,3** for short.

Sodium

The table of elements on page 32 lists sodium as $^{23}_{11}\mathbf{Na}$. From this we know its atomic number is **11**. This means it has 11 protons and 11 electrons.

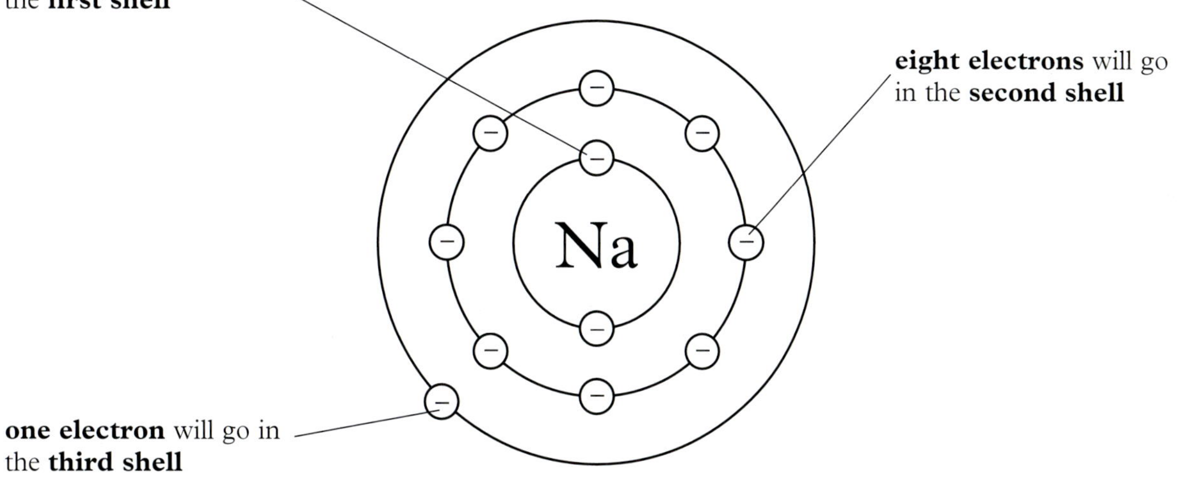
two electrons will go in the **first shell**
eight electrons will go in the **second shell**
Na
one electron will go in the **third shell**

Sodium has two electrons in the first shell, eight electrons in the second shell and one electron in the third shell. You can write this arrangement as **2,8,1** for short.

Questions

1 Use the Periodic Table on page 32 to help you.
a) Draw the electron shells for carbon (C), nitrogen (N), oxygen (O), fluorine (F) and neon (Ne).
b) Write the electron arrangements for each element in the shortened form.

2 Which elements have the following electron arrangements?
a) 2,2 b) 2,8,2 c) 2,4
d) 2,8,4 e) 2,8,7 f) 2,7

3 a) Which element has the atomic number 13?
b) For this element draw the electrons in their shells.
c) Write the electron arrangements in the shortened form.

4 a) Which element has atomic number 20?
b) For this element draw the electrons in their shells.
c) Write the electron arrangements in the shortened form for this element.

Electron arrangements and the Periodic Table

The Periodic Table is a list of all the elements in order of their atomic numbers. It is arranged in groups. Elements in the same group in the Periodic Table all have the same number of electrons in their outer shell except Group 0. Group 0 has different numbers because the outer shell is full. He has 2 electrons, Ne has 8 and Ar has 8. A shell with 2 or 8 electrons is full.

The number of electrons in the outer shell is the same as the number of the group, except for Group 0 which has full shells.

Group I	II	III	IV	V	VI	VII	0
H: **1**							He: 2
Li: 2,**1**	Be: 2,**2**	B: 2,**3**	C: 2,**4**	N: 2,**5**	O: 2,**6**	F: 2,7	Ne: 2,8
Na: 2,8,**1**	Mg: 2,8,**2**	Al: 2,8,**3**	Si: 2,8,**4**	P: 2,8,**5**	S: 2,8,**6**	Cl: 2,8,7	Ar: 2,8,8

Summary

An atom has a nucleus in its centre and electrons moving round the nucleus. The two types of particles found in the nucleus are protons and neutrons. The mass of a proton = the mass of a neutron = 1 atomic mass unit. Electrons have almost no mass.

Protons have a charge of (+1), neutrons have no charge, electrons have a charge of (–1). In an atom the number of protons = the number of electrons. Also in an atom, the number of neutrons = the mass number – the atomic number. Isotopes are atoms of the same element with different numbers of neutrons.

Electrons move round the nucleus in shells. The closest shell to the nucleus is called the first shell. It holds up to two electrons. The second shell holds up to eight electrons. The third shell holds up to eight electrons. The number of electrons in the outer shell of an atom is the same as the number of its Periodic Table group.

Key words

atomic number The number of protons in the nucleus of an atom of an element.

electron arrangement This is a quick way of telling us where electrons are found in an atom. For example 2,8 means two electrons in the first shell and eight electrons in the next shell.

mass number The number of protons + number of neutrons in the nucleus of one atom of an element.

End of Chapter 3 questions

1 The diagram shows the atomic structure of an atom of helium $^{4}_{2}He$.

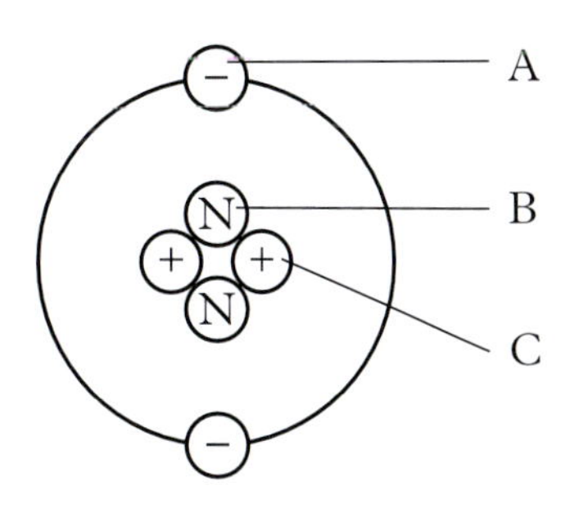
A
B
C
N
N
+
+
–
–

a) i) Name particle A.
ii) Name particle B.
iii) Name particle C.

b) What is the atomic number of helium?

2 a) Hydrogen has an atomic number of 1. Write down three things that this tells you about an atom of hydrogen.

b) The following diagrams show the electron arrangements of the atoms of three elements.

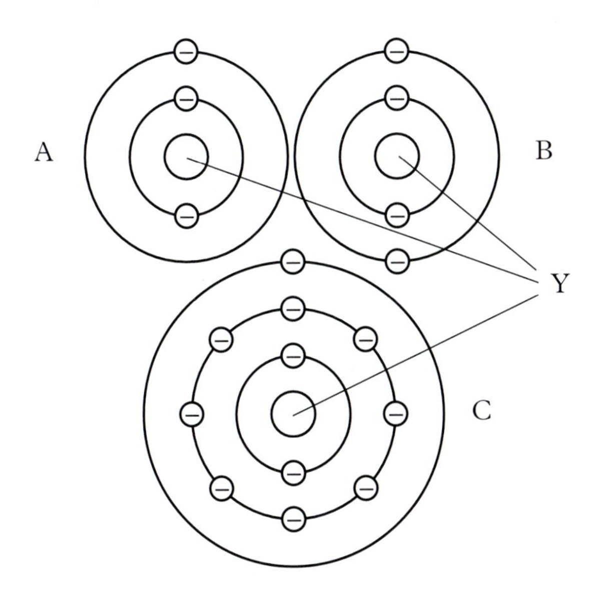
A
B
C
Y

c) Name the part of the atoms labelled Y.

d) Which two elements are in the same group in the Periodic Table? Give a reason for your answer.

3 Use the table of elements on page 32 to help you answer the following questions about an atom of sodium:
a) What is the atomic number of sodium?
b) What is the mass number of sodium?
c) How many protons does an atom of sodium have?
d) How many neutrons does an atom of sodium have?
e) How many electrons does an atom of sodium have?
f) Where are the protons and the neutrons found in an atom of sodium?

4 An atom of chlorine has 17 electrons.
Copy and complete the diagram below to show how these electrons are arranged in shells.

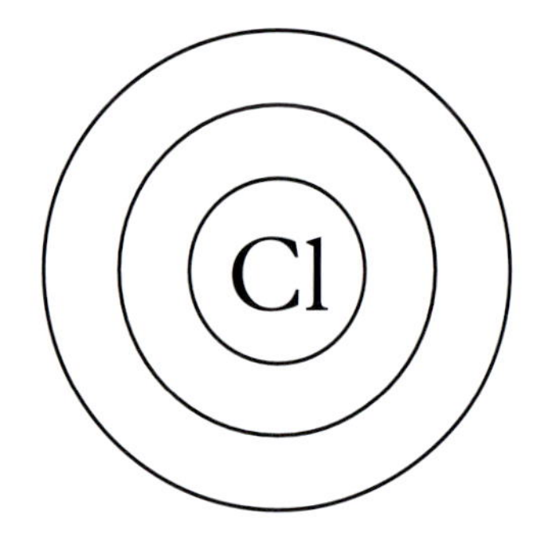
Cl

5 An atom of the element boron contains the following number of particles: five protons, five electrons and six neutrons.
a) What is boron's atomic number.
b) What is boron's mass number?
c) What is the charge on a proton?
d) What is the charge on a neutron?
e) Explain why the number of electrons and the number of protons in an atom of boron are equal.
f) How are boron's five electrons arranged in shells? You may answer using words, a diagram or shorthand notation.

6 The table below shows the number of particles which make up some atoms.

Element	Protons	Neutrons	Electrons
A	2	2	2
B	6	6	6
C	6	7	6
D	9	10	10

a) What is the mass number of element A?
b) Which two letters in the table represent a pair of isotopes?
c) Which two atoms have full outer shells of electrons?

4 How atoms bond together

Without chemical bonds we and most things around us wouldn't exist. All materials would be gases made up of single atoms. Materials are built from millions of atoms. Solid materials hold their shape because of the bonds between the atoms. Chemical bonds glue atoms together.

You can think of chemical bonds as acting like the holes and bumps that let you build things from model bricks. Each brick represents an atom.

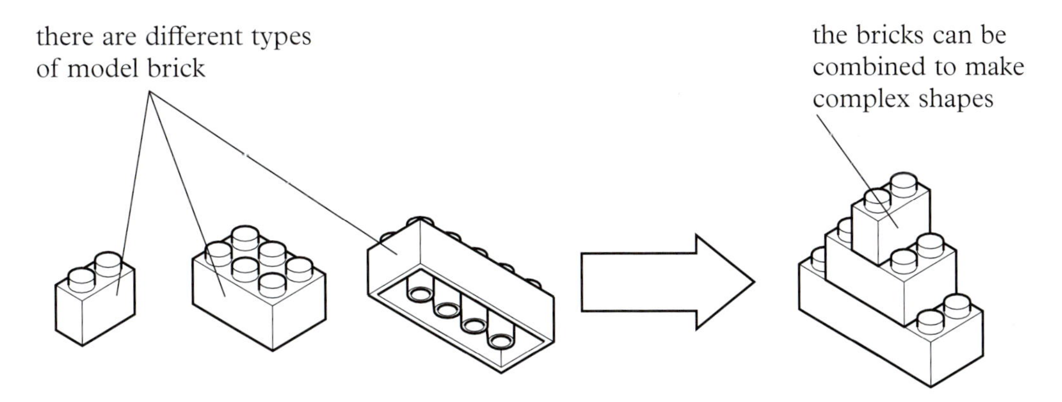
there are different types of model brick
the bricks can be combined to make complex shapes

Electrons

Electrons are tiny negatively-charged particles that orbit an atom's nucleus. Electrons move round the nucleus in **shells** that are different distances from the nucleus.

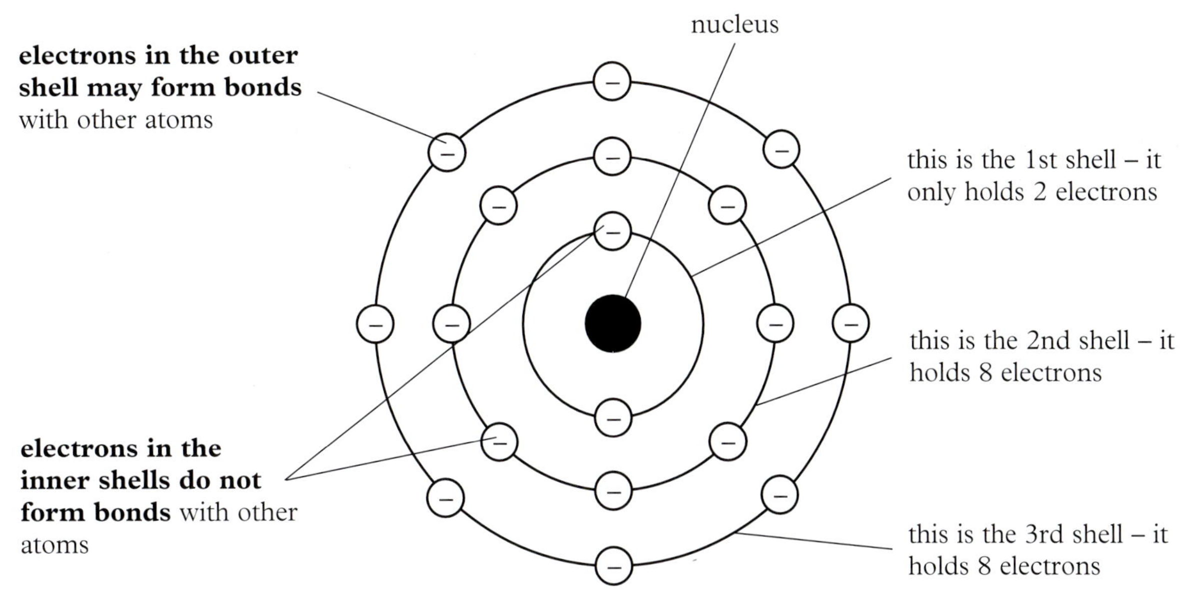
nucleus
electrons in the outer shell may form bonds with other atoms
this is the 1st shell – it only holds 2 electrons
this is the 2nd shell – it holds 8 electrons
electrons in the inner shells do not form bonds with other atoms
this is the 3rd shell – it holds 8 electrons

The electrons in the outer shell are the ones that form bonds with other atoms because they are on the outside of the atom.

The inert gases

There are six inert gases: helium, neon, argon, krypton, xenon and radon. These six gases are the only elements that always go around as single atoms. They are called the **inert** or **noble** gases because they do not react or combine with other elements.

inert gases exist as single atoms

inert gases do not form bonds

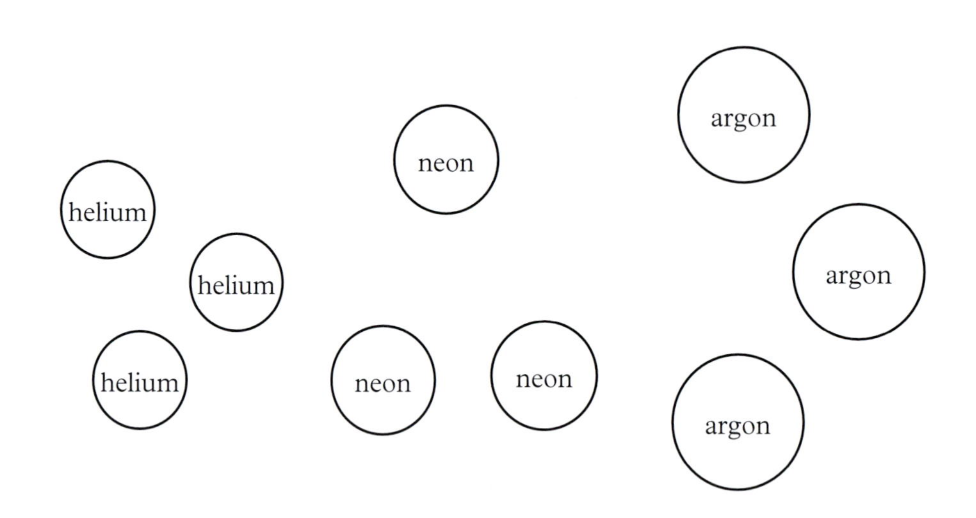
helium
helium
helium
neon
neon
neon
argon
argon
argon

The inert gases are the only elements which have full outer shells of electrons.

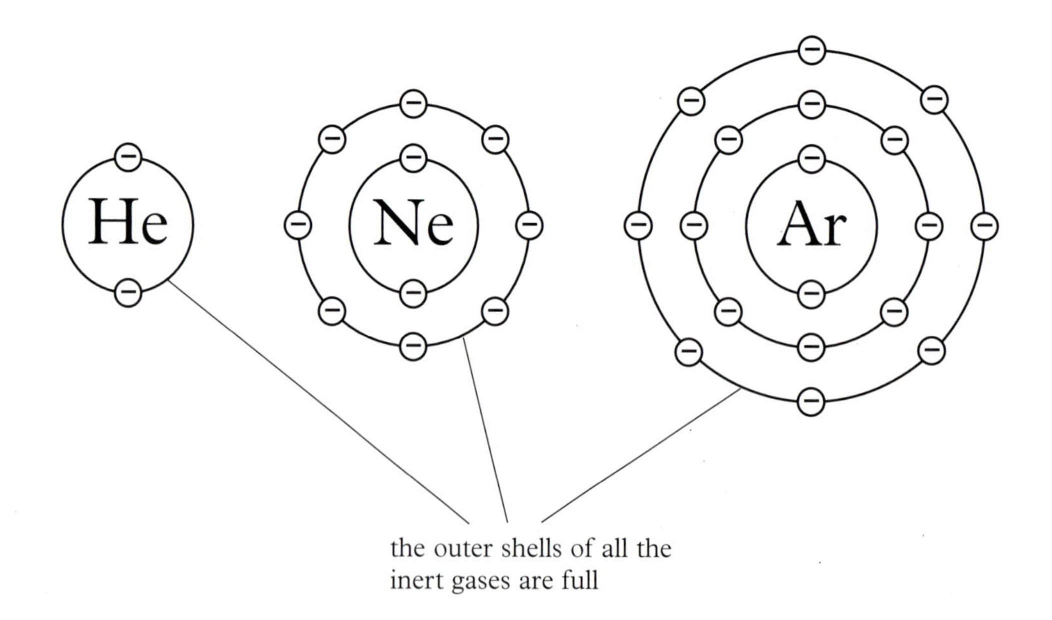
He
Ne
Ar
the outer shells of all the inert gases are full

The number of electrons in the outer shell

Inert gases have full outer shells of electrons. The atoms of all other elements have outer shells that are not full.

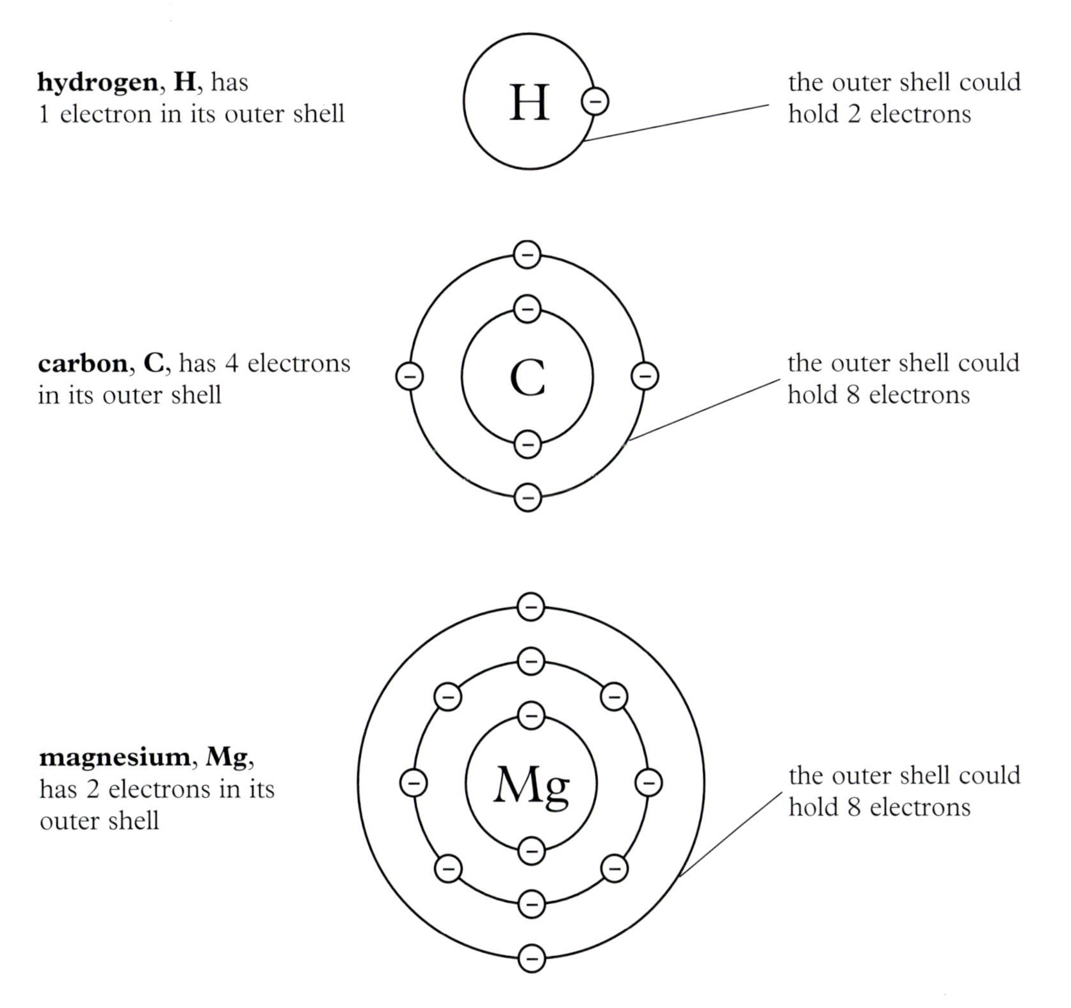

hydrogen, H, has
1 electron in its outer shell
H
the outer shell could
hold 2 electrons
carbon, C, has 4 electrons
in its outer shell
C
the outer shell could
hold 8 electrons
magnesium, Mg,
has 2 electrons in its
outer shell
Mg
the outer shell could
hold 8 electrons

It may help you to understand bonding if you imagine that all atoms 'want' a full outer shell of electrons. The inert gases have atoms with full outer shells of electrons. This explains why the atoms of inert gases do not form bonds with other atoms.

The atoms of all the other elements are always bonded to other atoms.

Atoms bond together so that they have full outer shells of electrons. There are three main ways by which atoms can get a full outer shell of electrons

ionic bonding
covalent bonding
metallic bonding.

Ionic bonding – the outer electrons move from one atom to another

When metals and non-metals bond together they form **ionic bonds**. To form an ionic bond, electrons are given away from the metal atoms and taken in by the non-metal atoms. We use **dot-cross diagrams** to show bonding. The dots and crosses both represent electrons which are identical, but using dots and crosses makes it clearer to see which electrons belong to which atom.

For example, the metal lithium and the non-metal fluorine bond together:

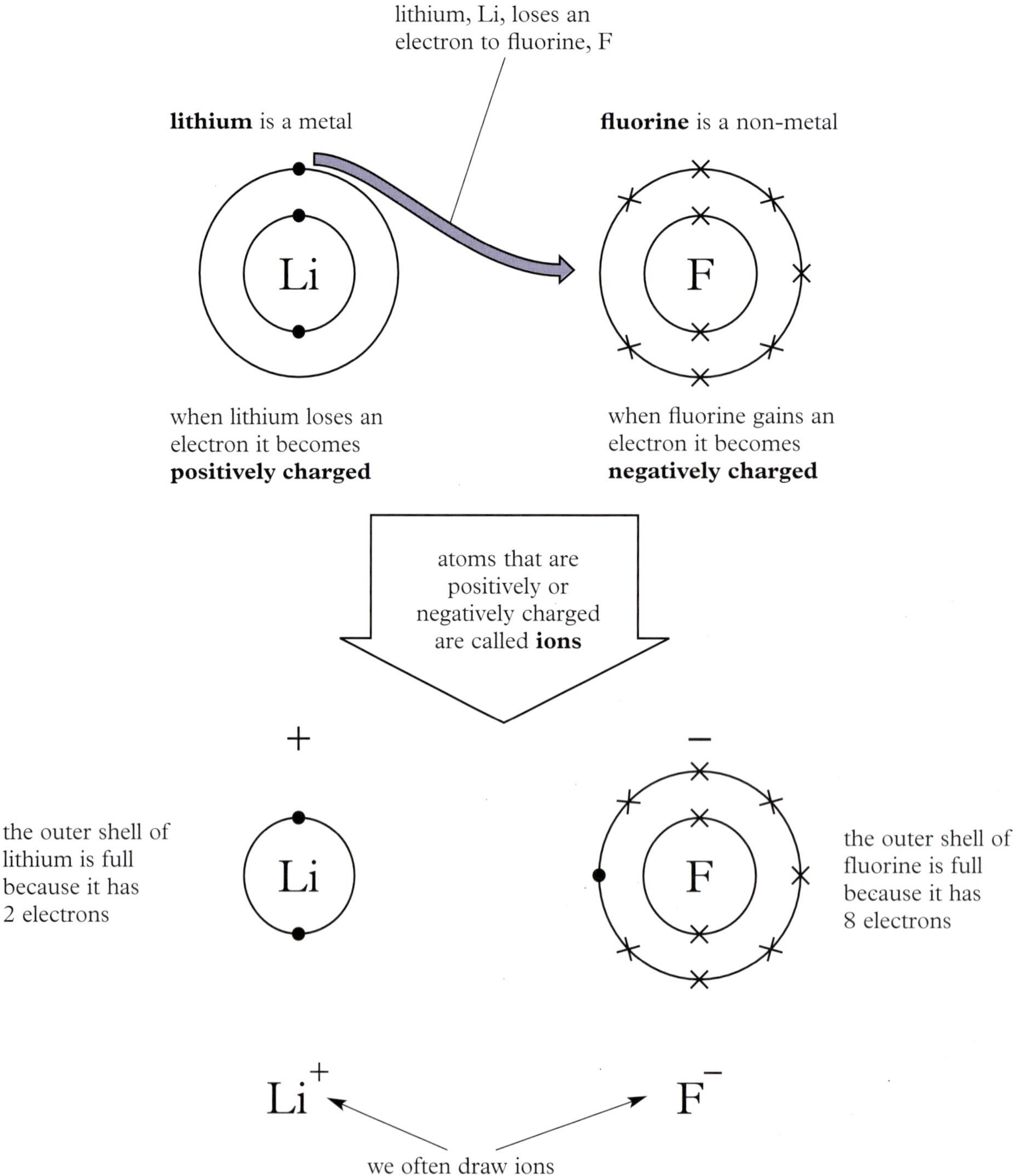
lithium, Li, loses an electron to fluorine, F
lithium is a metal
fluorine is a non-metal
Li
F
when lithium loses an electron it becomes **positively charged**
when fluorine gains an electron it becomes **negatively charged**
atoms that are positively or negatively charged are called **ions**
+
–
the outer shell of lithium is full because it has 2 electrons
Li
F
the outer shell of fluorine is full because it has 8 electrons
Li^+
F^-
we often draw ions like this

How ions form a compound

The ions of lithium and fluorine form the solid compound lithium fluoride:

these ions of lithium and fluorine are held together by the attraction of positive charges to negative charges

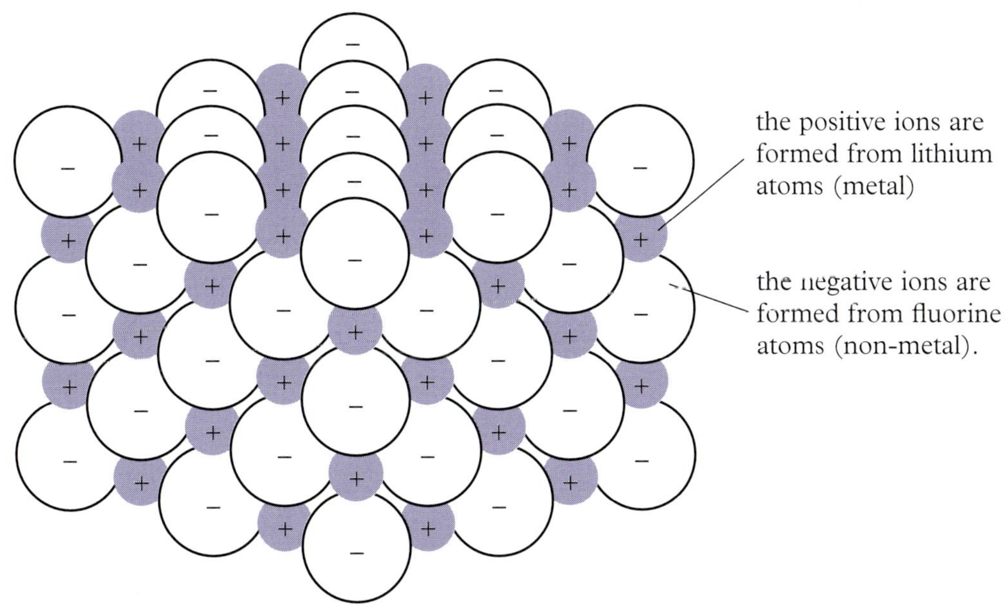
the positive ions are formed from lithium atoms (metal)
the negative ions are formed from fluorine atoms (non-metal).

the compound has no overall charge because there is the same amount of positive charge as negative charge

this kind of structure, where the bonding continues throughout, is called a **giant structure**

Ionic bonds are strong and spread throughout the compound. This means that ionic compounds, like all giant structures, are solids which are difficult to melt. Ionic compounds have high melting points.

Questions

1 In an ionic compound which sort of ion is positively charged?

2 Which sort of ion is negatively charged?

3 What holds the ions together?

4 Why does an ionic compound have a giant structure?

5 Why is there no charge on an ionic compound, even though it is made up of charged particles?

Covalent bonding – the outer electrons are shared

Covalent bonds are formed between the atoms of non-metals. Materials which have covalent bonding may be:

- gases, liquids or solids which are easy to melt. These are made of molecules.
- solids which are hard to melt. These have giant structures.

Molecules

Fluorine gas, F_2, is made up of pairs of fluorine atoms:

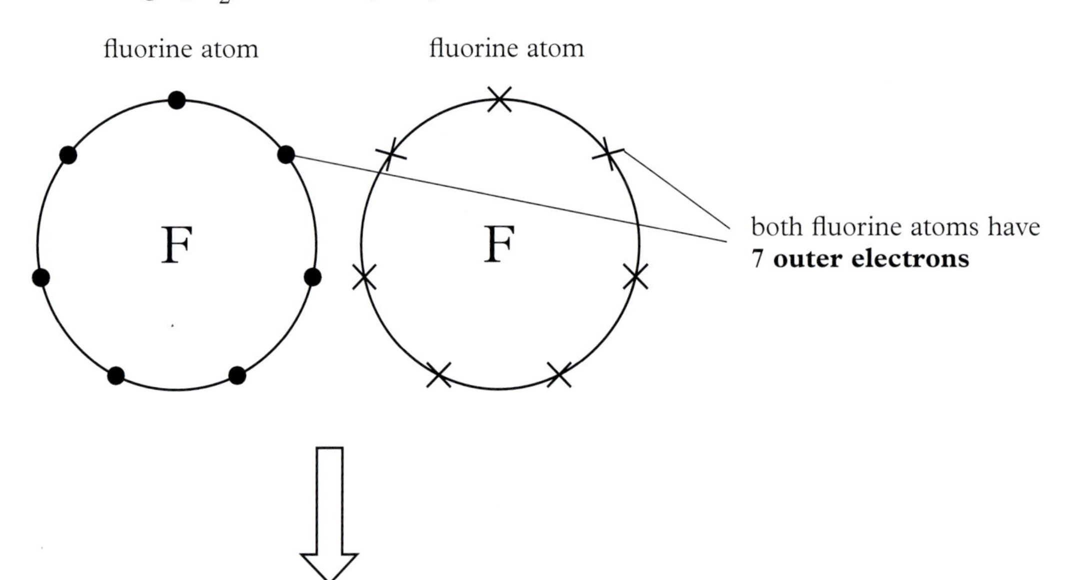
fluorine atom
fluorine atom
F
F
both fluorine atoms have
7 outer electrons

the two atoms share two electrons between them

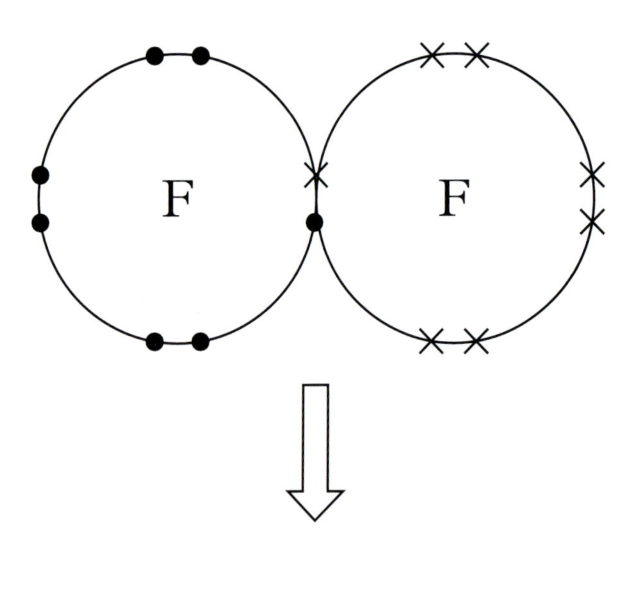
F
F

each atom now has 8 electrons

the sharing of electrons like this is called a **covalent bond**

a pair of fluorine atoms joined like this is called a fluorine **molecule**

F – F

we represent this molecule in a shorthand form where the dash is used to mean a covalent bond

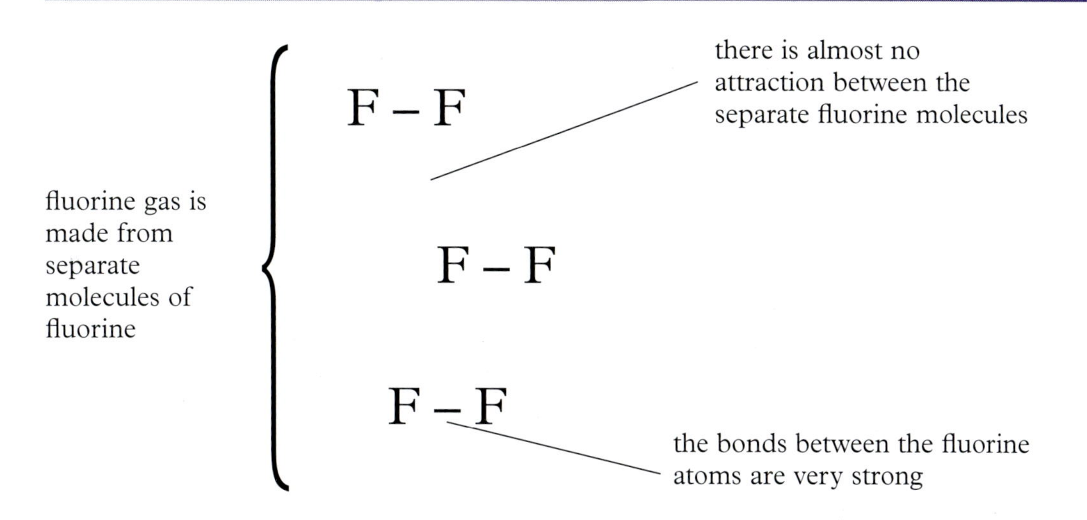
F – F
there is almost no attraction between the separate fluorine molecules
fluorine gas is made from separate molecules of fluorine
F – F
F – F
the bonds between the fluorine atoms are very strong

Fluorine, like all gases and liquids has a **molecular structure**.

Giant covalent structures

Diamond is an example of a giant covalent structure. It is made of carbon atoms.

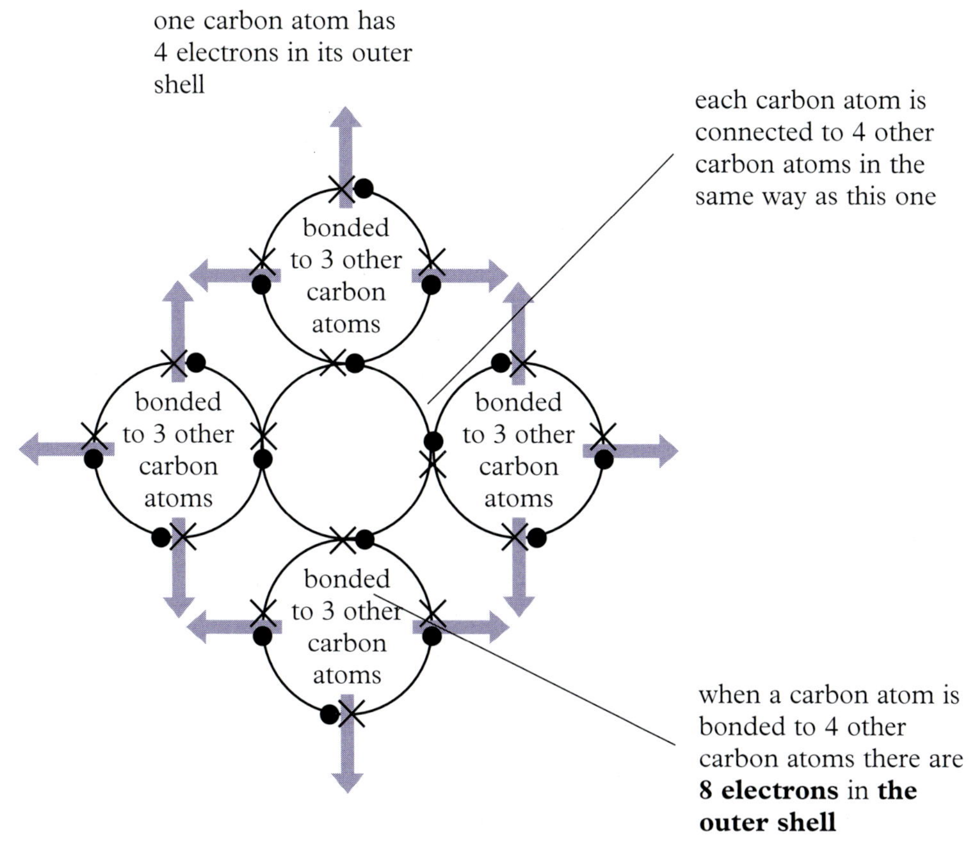
one carbon atom has 4 electrons in its outer shell
each carbon atom is connected to 4 other carbon atoms in the same way as this one
bonded to 3 other carbon atoms
bonded to 3 other carbon atoms
bonded to 3 other carbon atoms
bonded to 3 other carbon atoms
when a carbon atom is bonded to 4 other carbon atoms there are **8 electrons** in **the outer shell**

Using a dot-and-cross diagram to represent diamond becomes very complicated because of the number of covalent bonds. Instead we can represent a diamond molecule in the following way:

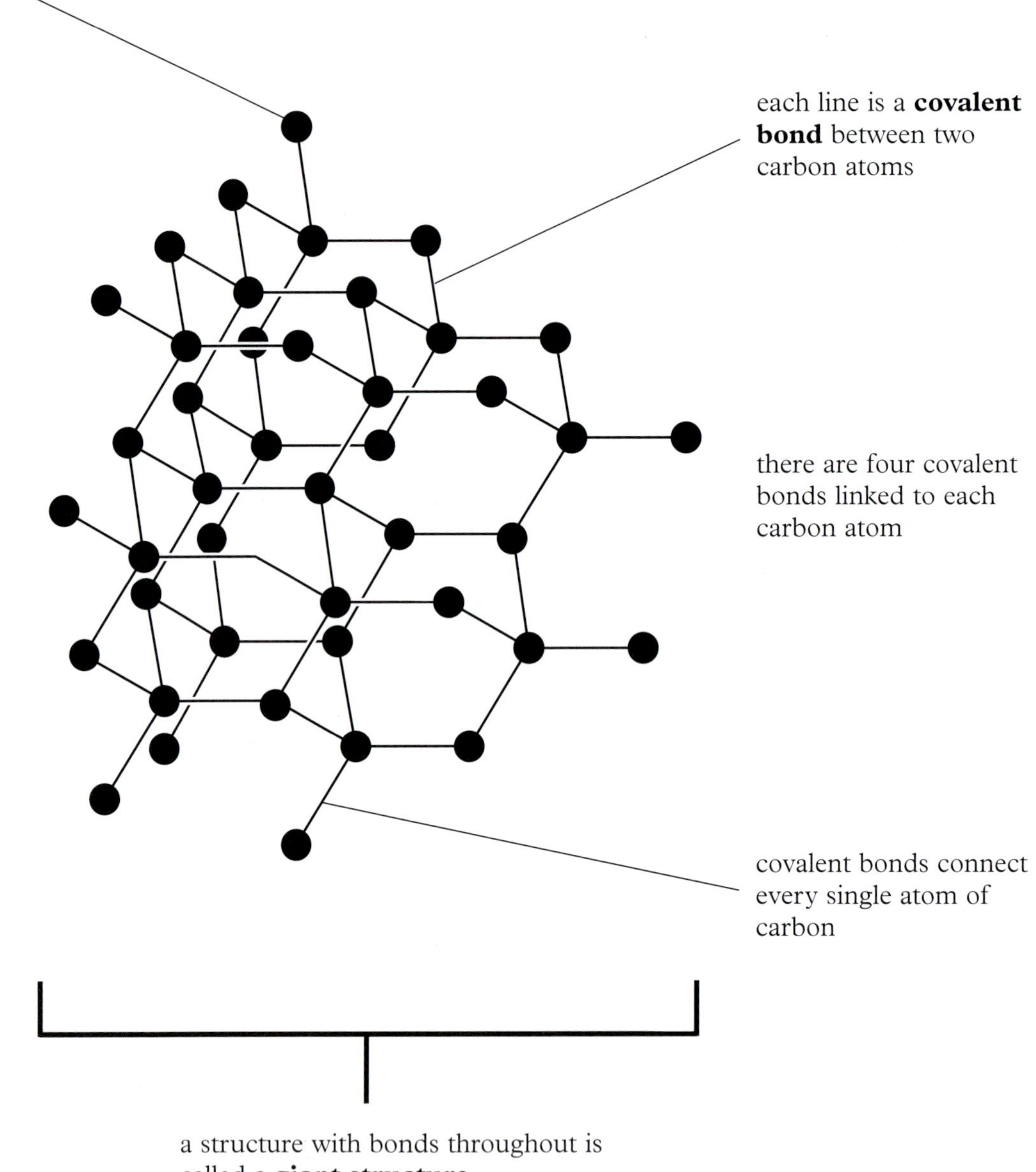
each sphere is a **carbon atom**
each line is a **covalent bond** between two carbon atoms
there are four covalent bonds linked to each carbon atom
covalent bonds connect every single atom of carbon
a structure with bonds throughout is called a **giant structure**

The strength of the covalent bonds and the large number of them means diamond is difficult to melt. The same is true for all giant structures.

Metallic bonding

Metallic bonding happens when metal atoms give up one or two electrons from their outer shells. Any material with metallic bonding is called a **metal**. All metals have giant structures.

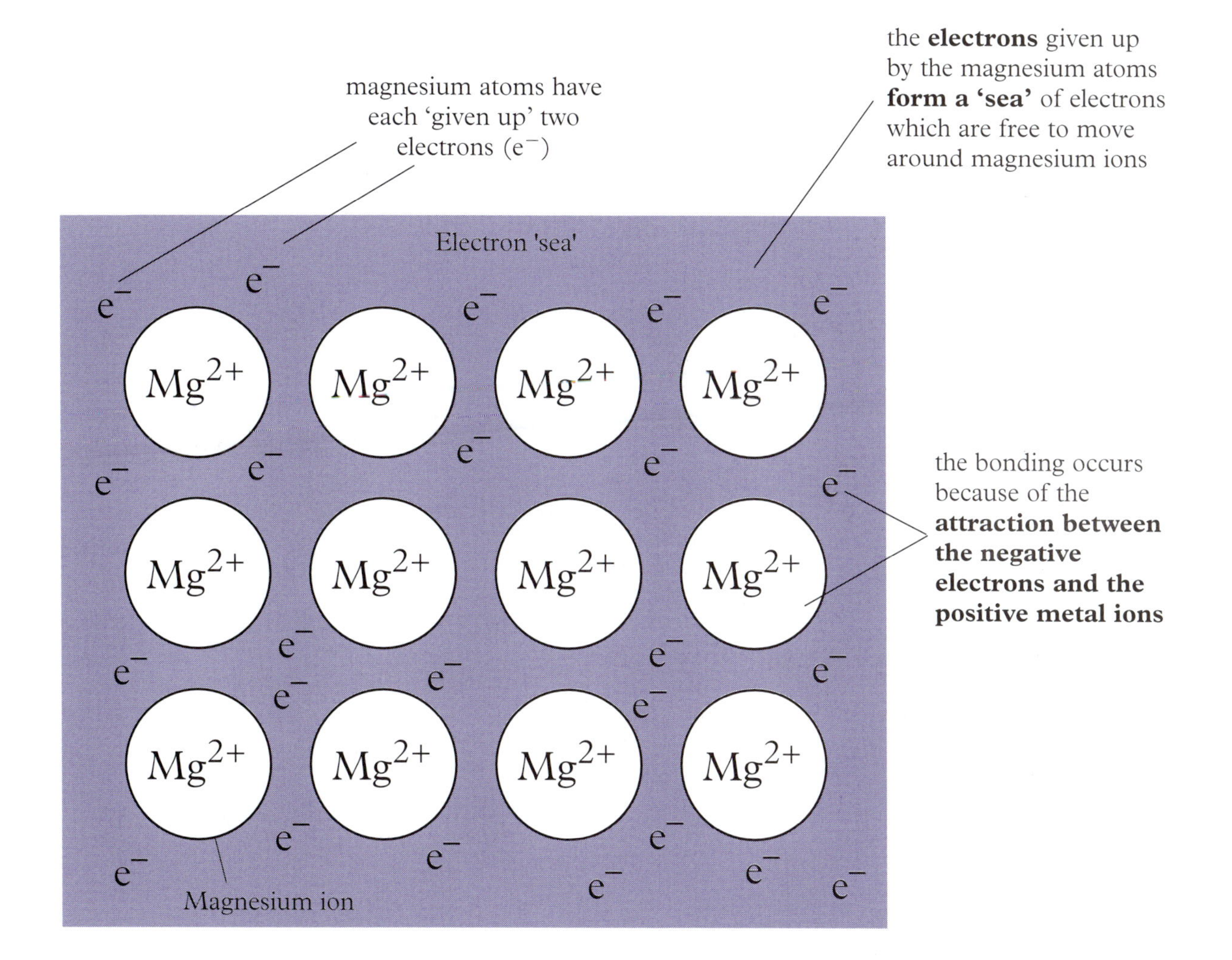
magnesium atoms have each 'given up' two electrons (e^-)
the **electrons** given up by the magnesium atoms **form a 'sea'** of electrons which are free to move around magnesium ions
Electron 'sea'
Mg^{2+}
e^-
the bonding occurs because of the **attraction between the negative electrons and the positive metal ions**
Magnesium ion

Questions

1 From the names of the compounds below, decide which three have ionic bonding and say why you made your choice. (HINT: remember that ionic bonds form between metals and non-metals).

a) copper oxide
b) carbon dioxide
c) magnesium oxide
d) carbon monoxide
e) sodium chloride

2 If something is 'difficult to melt', will it have a high melting point or a low melting point?

How different substances behave

The type of bonding in a substance affects how it behaves (its **properties**) – for example its melting point, boiling point and how it conducts electricity.

Compounds with ionic bonding

Structures of ionic compounds

Ionic compounds always have giant structures.

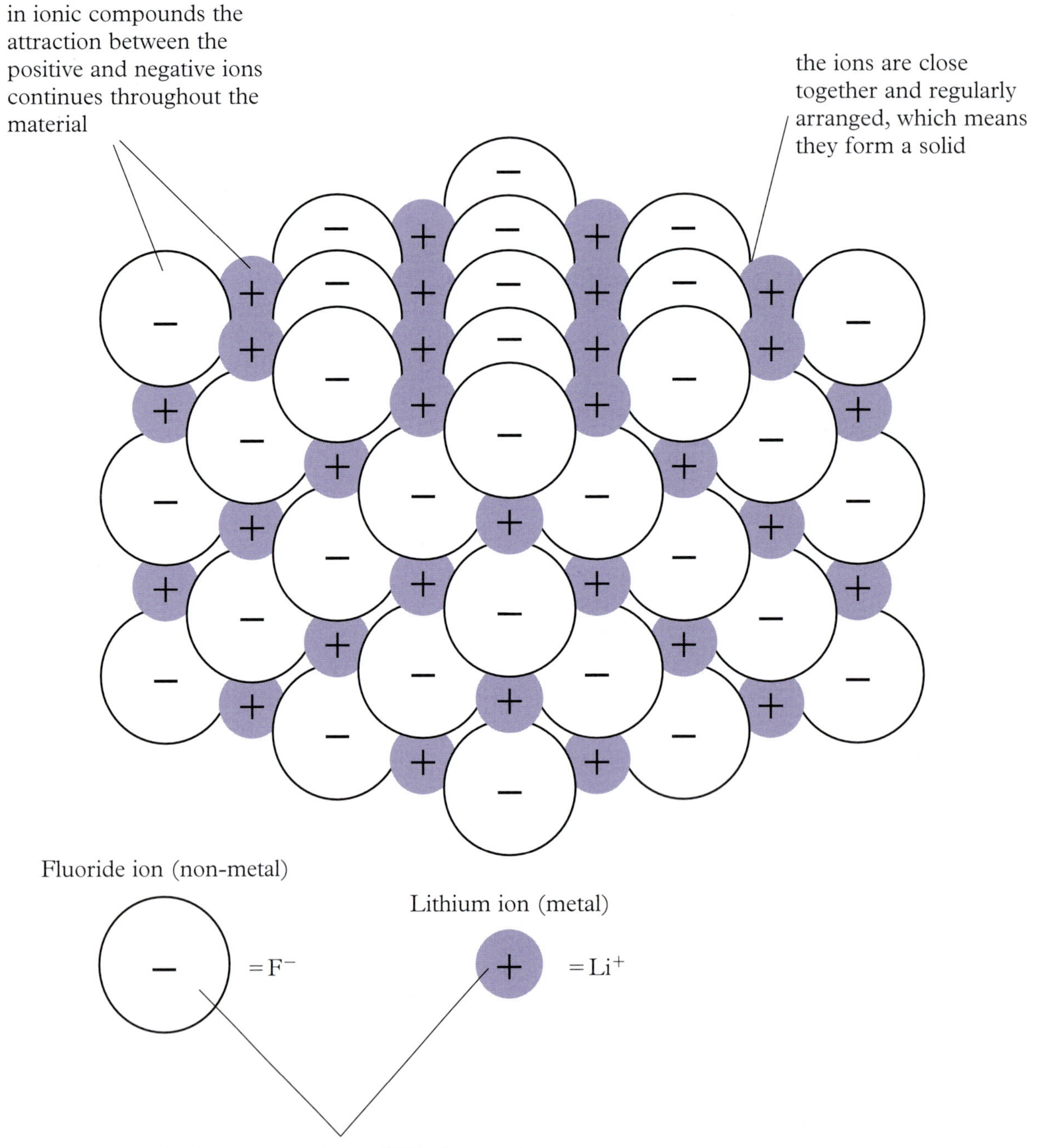
in ionic compounds the attraction between the positive and negative ions continues throughout the material
the ions are close together and regularly arranged, which means they form a solid
Fluoride ion (non-metal)
$=F^-$
Lithium ion (metal)
$=Li^+$
ionic compounds are difficult to melt because it is hard to separate the ions

Conducting electricity through ionic compounds

Ions are charged particles. The movement of charged particles forms an electric current. Materials made up of charged particles which are free to move can conduct electricity.

Ionic compounds will not conduct electricity when solid.

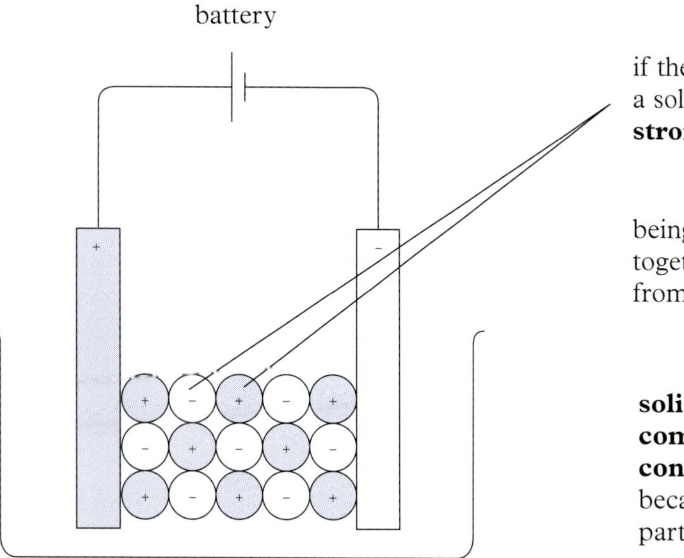
battery
+
-

if the ionic compound is a solid then the **ions are strongly held together**

being strongly held together stops the ions from moving around

solid ionic compounds will not conduct electricity because the charged particles are not free to move

How ionic compounds conduct electricity

Ionic compounds will conduct electricity when they are dissolved in water, or when they are melted to a liquid.

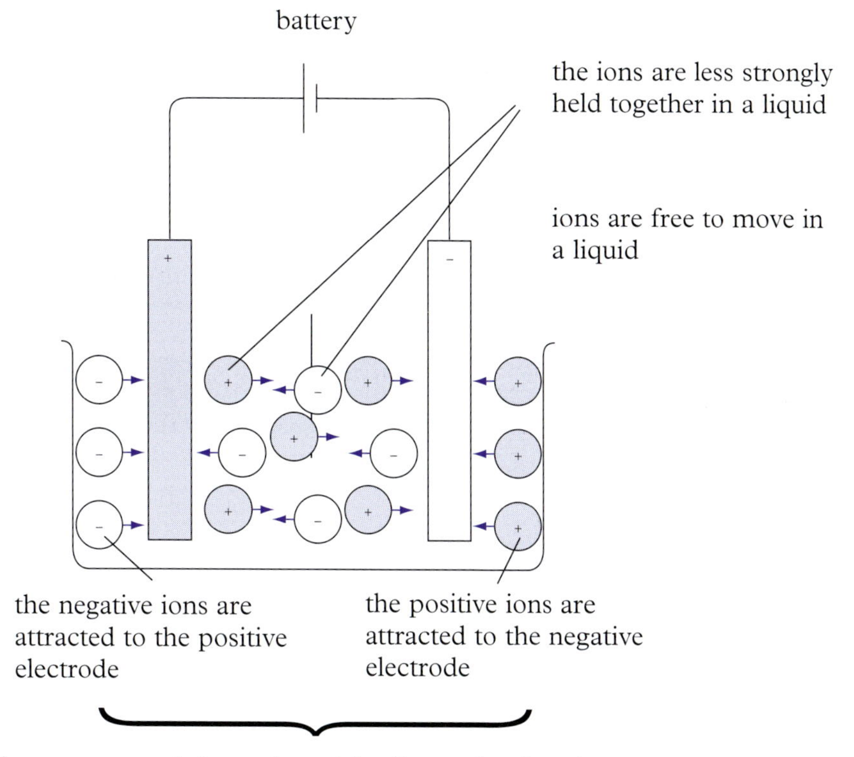
battery
+
-
the ions are less strongly held together in a liquid
ions are free to move in a liquid
the negative ions are attracted to the positive electrode
the positive ions are attracted to the negative electrode

this movement of charged particles forms the electric current.

Covalently bonded substances

Substances with covalent bonds may form simple molecular structures or giant structures.

Structure

1 **Molecular structures**
Covalently bonded substances made up of molecules form gases, liquids or solids with low melting points.

2 **Giant structures**
The bonding extends throughout giant structures. The bonding makes substances with giant structures form solids which are hard to melt.

Conducting electricity

Covalently bonded structures have no charged particles so they will not conduct electricity either as solids, liquids or in solutions.

Metals

Structure

Metals have a giant structure because the bonding extends throughout. They are solids which are hard to melt.

Conducting electricity

Metals have a 'sea' of electrons which are free to move. This movement of electrons means that metals can conduct electricity.

How to tell the difference

By looking at the way in which any material behaves you can usually tell whether the structure is giant or molecular and whether the bonding is ionic, molecular or covalent.

The differences between ionic, covalent and metallic bonding

Compounds with ionic bonding often dissolve in water. The table below gives you a summary of the important differences between substances with ionic, covalent and metallic bonding.

Type of elements present	Does it conduct electricity?			Bonding	Is it easy to melt?	Structure
	solid	liquid	(aq)*			
metal and non-metal	No	Yes	Yes	Ionic	No	Giant
non-metal and non-metal	No	No	No	Covalent	No	Giant
non-metal and non-metal	No	No	No	Covalent	Yes	Molecular
metal and metal	Yes	Yes	Yes	Metallic	No	Giant

*(aq) means dissolved in water

1 You need to find out if the material is a gas or a liquid or a solid which melts easily.

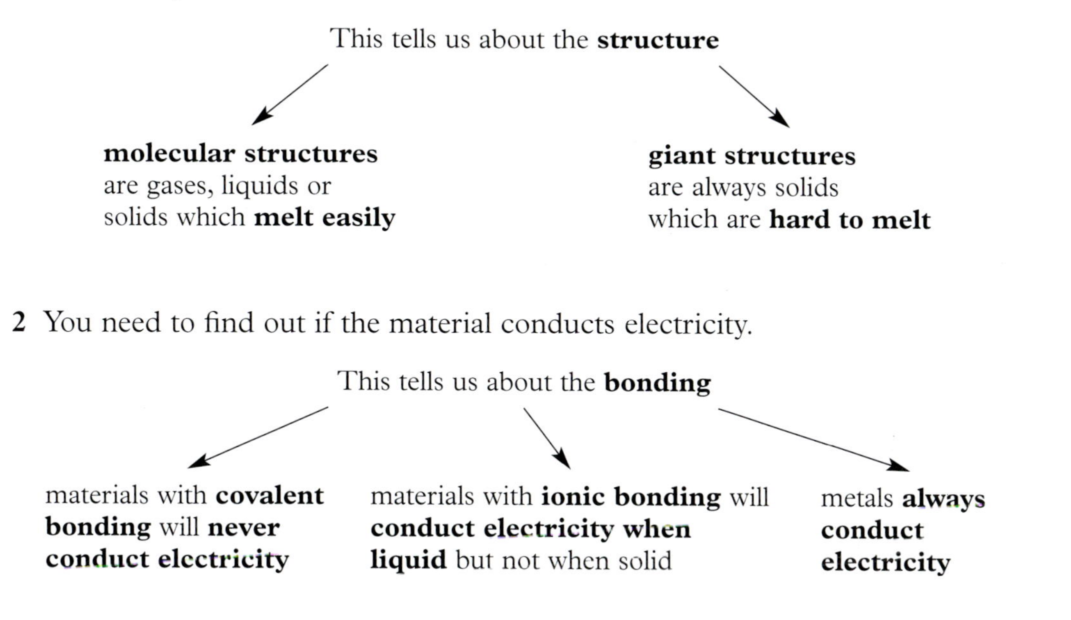
This tells us about the **structure**
molecular structures are gases, liquids or solids which **melt easily**
giant structures are always solids which are **hard to melt**
This tells us about the **bonding**
materials with **covalent bonding** will **never conduct electricity**
materials with **ionic bonding** will **conduct electricity when liquid** but not when solid
metals **always conduct electricity**

2 You need to find out if the material conducts electricity.

Questions

1 From the descriptions of the substances below say which three are made of **molecules** and say why you made your choice:

a) water – a colourless liquid.
b) marble – a hard white rock which does not melt easily.
c) candle wax – a white solid which is easy to melt.
d) sand – a white solid which does not melt easily.
e) methane – a colourless gas.

2 a) You have a pot of sugar and a pot of salt which look very similar. Sugar is made up of covalently bonded molecules whereas salt is an ionic compound. Without tasting, how could you test to find out which is salt and which is sugar?

b) You have a pot of fine white sand and a pot of salt which look very similar. They both have giant structures but sand is covalently bonded whereas salt has ionic bonds. Without tasting, how could you test to find out which is salt and which is sand?

Summary

Chemical bonds are forces that hold atoms together. There are three kinds of chemical bonds, **ionic bonds**, **covalent bonds** and **metallic bonds**.

Ionic bonds occur when a metal transfers electrons to a non-metal. The resulting positive metal ions and negative non-metal ions form a giant structure. Ionic compounds are hard to melt. Ionic compounds conduct electricity only when melted or dissolved in water.

Covalent bonds occur when non-metals share electrons. Covalently bonded materials may be made up of molecules in the form of gases, liquids or solids that are easy to melt, or giant structures which are hard to melt. Covalently bonded substances do not conduct electricity at all.

Metallic bonds occur when metals give away some electrons to form a 'sea' of electrons. Metals have giant structures and are hard to melt. Metals always conduct electricity.

Structure describes how the atoms are arranged. There are two sorts of structures: giant structures, which have high melting points, and molecular structures, which have low melting points.

Key words

giant structure Solids in which the bonding extends throughout the material.

ions Particles with either a positive or a negative charge.

molecules Separate particles with no charge, with covalent bonds holding together the atoms within them.

End of Chapter 4 questions

1 Choose the best word from the list to fill each of the spaces. The words may be used more than once.

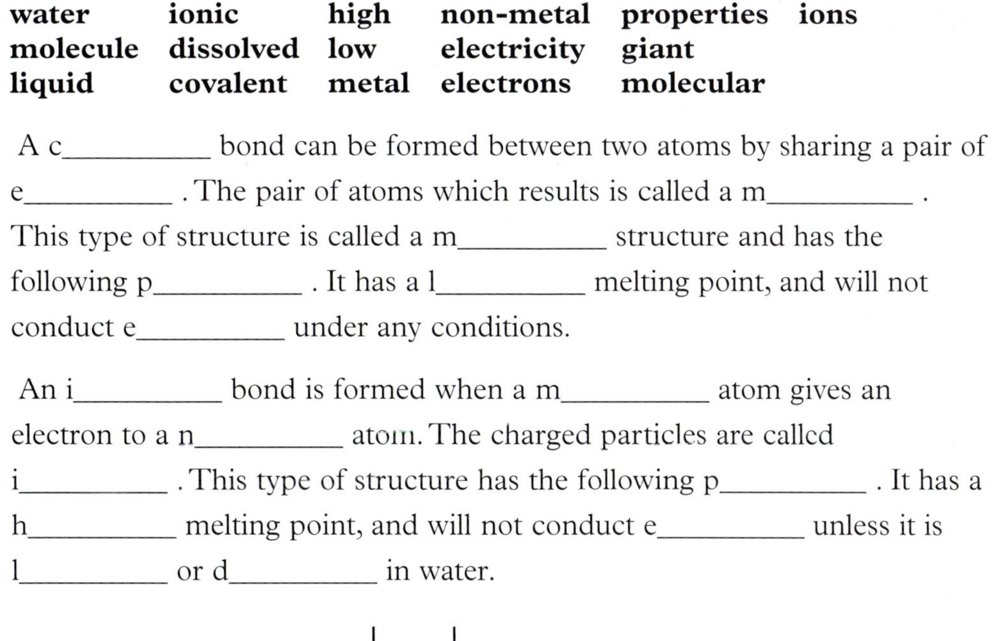

water	**ionic**	**high**	**non-metal**	**properties**	**ions**
molecule	**dissolved**	**low**	**electricity**	**giant**	
liquid	**covalent**	**metal**	**electrons**	**molecular**	

A c__________ bond can be formed between two atoms by sharing a pair of e__________ . The pair of atoms which results is called a m__________ . This type of structure is called a m__________ structure and has the following p__________ . It has a l__________ melting point, and will not conduct e__________ under any conditions.

An i__________ bond is formed when a m__________ atom gives an electron to a n__________ atom. The charged particles are called i__________ . This type of structure has the following p__________ . It has a h__________ melting point, and will not conduct e__________ unless it is l__________ or d__________ in water.

2

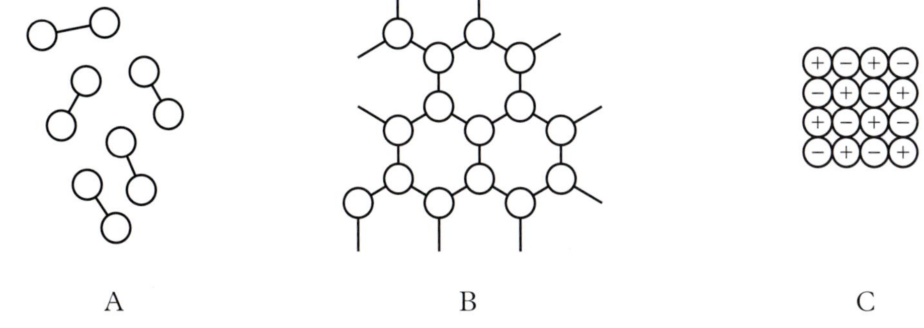
A
B
C

Look at the diagrams of chemical structures and then answer the following:

a) Which is made of molecules?
b) Which one has a giant ionic structure?
c) Which one is made of ions?
d) Which one is made up of a metal and a non-metal?
e) Which two will be hard to melt?
f) Which one is a gas?
g) Which one has a giant covalent structure?
h) Name a substance that has the same type of structure as A.
i) Name a substance that has a structure like B.
j) Give the name of a substance that has the same type of structure as C.

3 The table below gives some information about four substances W, X, Y and Z.

Substance	Melting point °C	Boiling point °C	Electrical conduction as a solid	Electrical conduction as a liquid
W	−95	56	poor	poor
X	801	1467	poor	good
Y	1083	2595	good	good
Z	1610	2230	poor	poor

a) Which substance is liquid at room temperature (25 °C)?
b) Which substance is a metal? Explain your answer.
c) Which substance is an ionically bonded compound? Give reasons for your answer.
d) Which compound is a giant covalently bonded compound? Give reasons for your answer.
e) Which compound consists of covalently bonded molecules? Give reasons for your answer.

4 a)

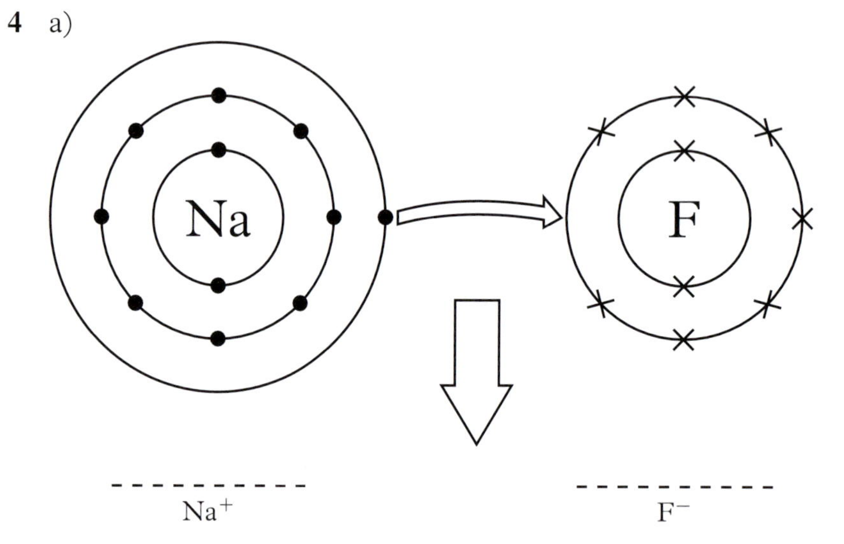
Na
F
Na+
F−

The electron arrangement of sodium, Na, is 2,8,1 and of fluorine, F, is 2,7.

Copy and complete the electron diagrams to explain how sodium and fluorine bond together to form sodium ions and fluoride ions.

5 Magnesium has the electron arrangement 2,8,2. Oxygen has the electron arrangement 2,6.
a) Draw electron diagrams for magnesium, Mg, and oxygen, O.
b) Show how magnesium loses its two outer electrons to oxygen to form a magnesium ion and an oxygen ion.
c) What is the name of the compound they form?

5 The masses of atoms and molecules

Atoms are very small indeed. For example, there are roughly 100 000 000 000 000 000 000 atoms of iron on a pinhead. This means that we can never take one atom and weigh it to find out how heavy it is. Scientists have to use roundabout methods to weigh atoms.

Once scientists found out that the element hydrogen was the lightest of all atoms, they said that the **relative atomic mass** ($\mathbf{A_r}$) of hydrogen was 1. Then the atoms of other elements were compared in mass to hydrogen, to find their relative atomic masses.

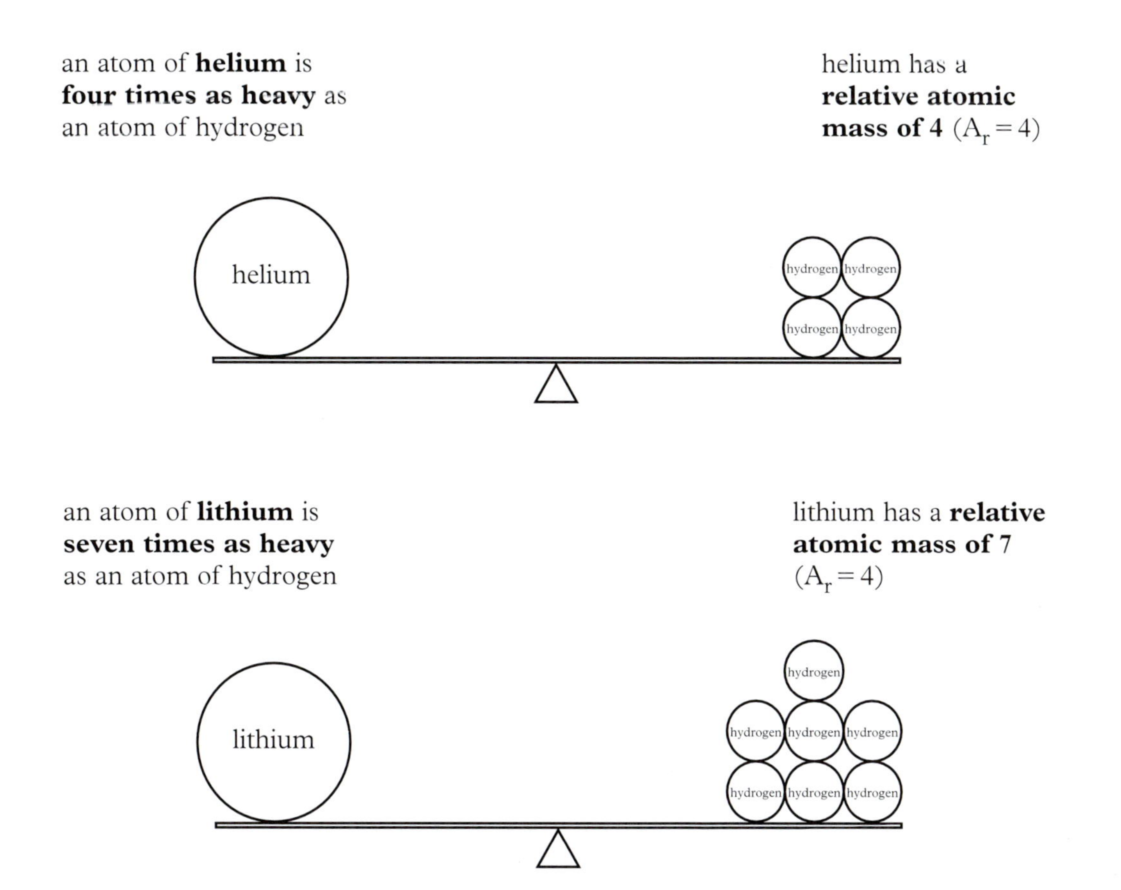
an atom of helium is four times as heavy as an atom of hydrogen
helium has a relative atomic mass of 4 ($A_r = 4$)
helium
hydrogen
hydrogen
hydrogen
hydrogen
an atom of lithium is seven times as heavy as an atom of hydrogen
lithium has a relative atomic mass of 7 ($A_r = 4$)
lithium
hydrogen
hydrogen
hydrogen
hydrogen
hydrogen
hydrogen
hydrogen

Every element can be written with two numbers, which you can find in the Periodic Table. One of these numbers is the relative atomic mass for that element.

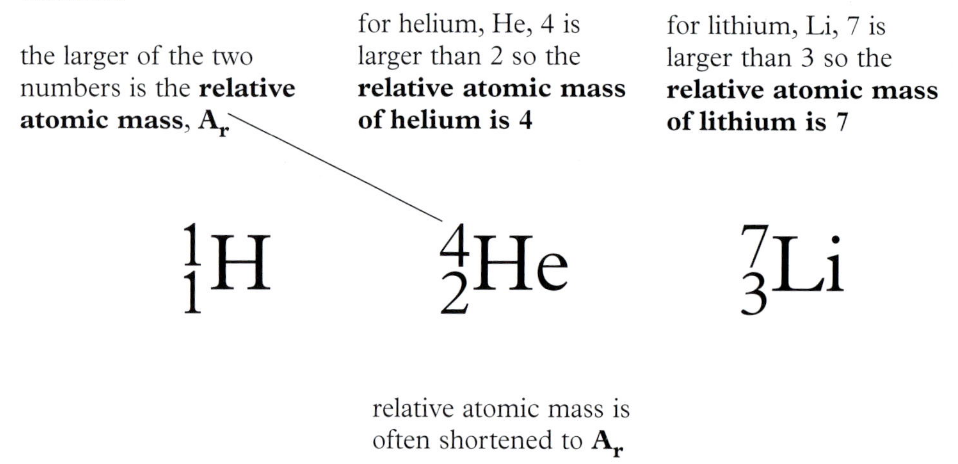
the larger of the two numbers is the **relative atomic mass, A_r**
for helium, He, 4 is larger than 2 so the **relative atomic mass of helium is 4**
for lithium, Li, 7 is larger than 3 so the **relative atomic mass of lithium is 7**
$^{1}_{1}H$
$^{4}_{2}He$
$^{7}_{3}Li$
relative atomic mass is often shortened to $\mathbf{A_r}$

The A_r is written at the top in some Periodic Tables and at the bottom in others but is always the larger number. The A_r is the average mass of an atom of the element compared with the mass of a hydrogen atom. It is just a number and has no units.

The relative formula mass

If you add together the A_rs of all the atoms in the formula of a substance, you have the **relative formula mass**, which is written as $\mathbf{M_r}$.

For example, the following elements have the following A_rs:
Mg = 24   O = 16   H = 1

Magnesium oxide

MgO

there is one magnesium atom, with an A_r of 24

there is one oxygen atom, with an A_r of 16

the M_r of magnesium oxide is **24 + 16 = 40**

Water

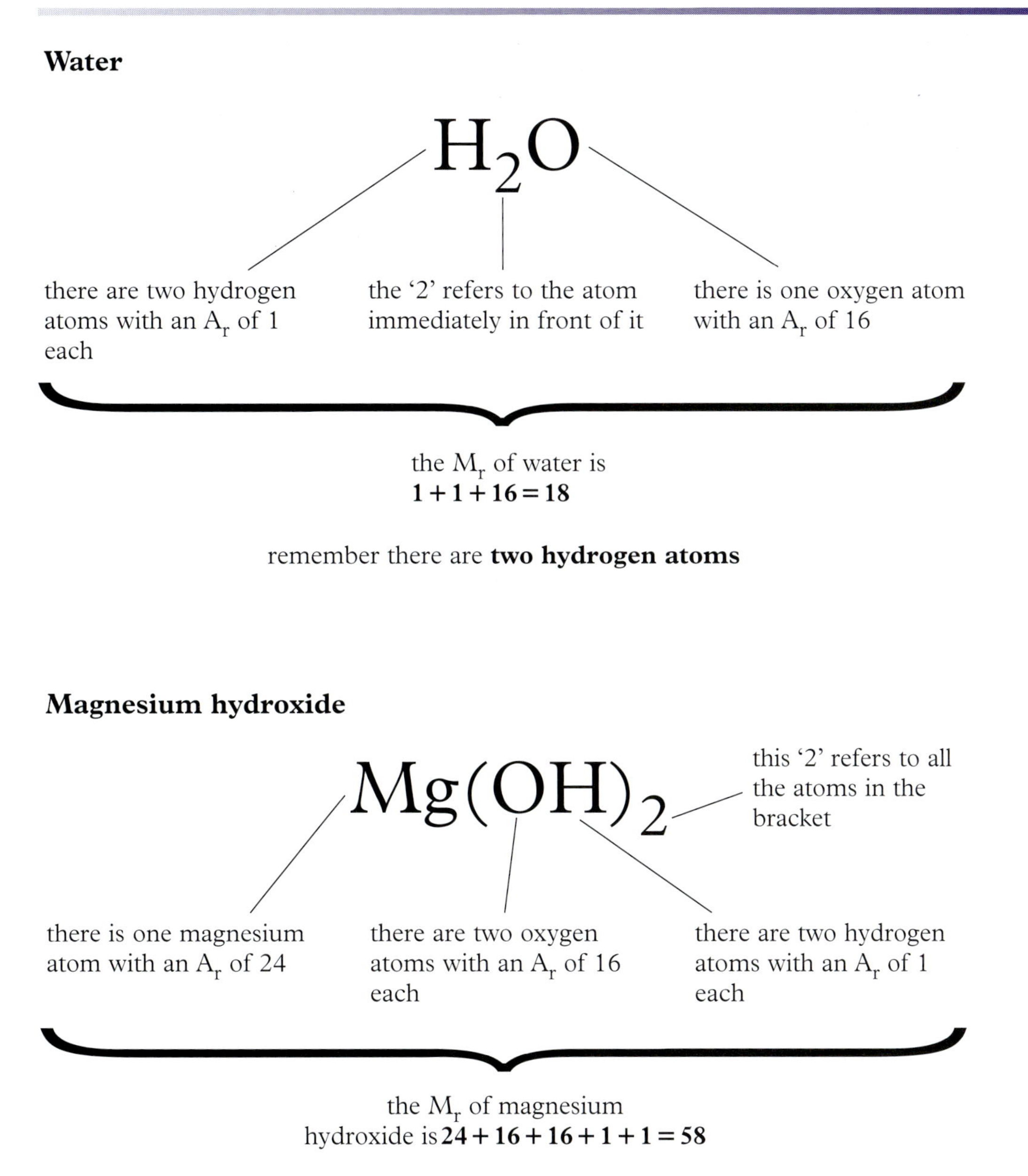

H_2O
there are two hydrogen atoms with an A_r of 1 each
the '2' refers to the atom immediately in front of it
there is one oxygen atom with an A_r of 16
the M_r of water is **1 + 1 + 16 = 18**
remember there are **two hydrogen atoms**
Magnesium hydroxide
$Mg(OH)_2$
this '2' refers to all the atoms in the bracket
there is one magnesium atom with an A_r of 24
there are two oxygen atoms with an A_r of 16 each
there are two hydrogen atoms with an A_r of 1 each
the M_r of magnesium hydroxide is **24 + 16 + 16 + 1 + 1 = 58**

Questions

1 Use the Periodic Table on page 185 to find the elements with the following A_rs:

a) 39
b) 12
c) 19
d) 28
e) 56

2 Find M_r for the following:

a) sodium chloride, NaCl
b) calcium carbonate, $CaCO_3$
c) sulphuric acid, H_2SO_4
d) ammonia, NH_3

The rules for chemical formulae

A small number after the symbol for an element tells us how many atoms of that element are present.

Water: H_2O means there are two H atoms and one O atom.

A small number after a bracket multiplies all the atoms inside the bracket.

Calcium nitrate: $Ca(NO_3)_2$ means there is one Ca atom and two × (one N and three O) = two N and six O atoms.

Using the relative formula mass

The relative formula mass can tell us the percentage of any element present in a compound.

Look at magnesium oxide which has the formula MgO. We know that A_r for Mg = 24 and O = 16.

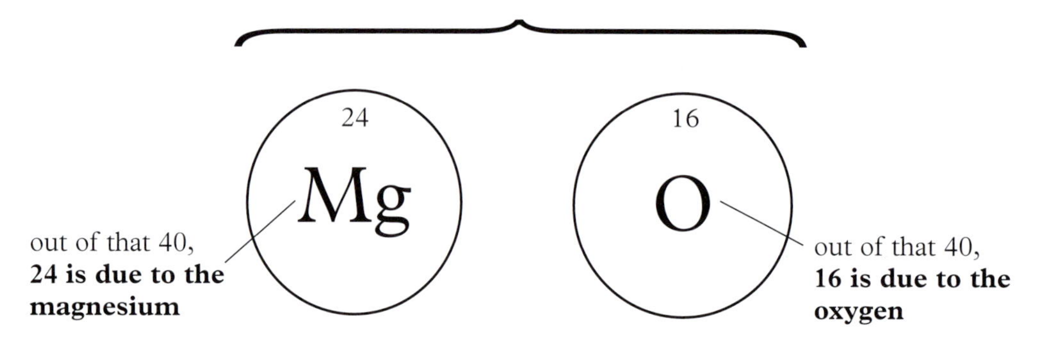
the M_r of magnesium oxide is 40
24
Mg
16
O
out of that 40, **24 is due to the magnesium**
out of that 40, **16 is due to the oxygen**

If we have 40 g of magnesium oxide, we know that 24 g of it is magnesium and 16 g of it is oxygen.

To find the percentage of magnesium present, we divide the mass of magnesium by the mass of magnesium oxide and multiply by 100.

$$\frac{\text{mass of magnesium}}{\text{mass of magnesium oxide}} \times 100\% = \frac{24\,\text{g}}{40\,\text{g}} \times 100\% = 60\%$$

To find the percentage of oxygen present, we divide the mass of oxygen by the mass of magnesium oxide and multiply by 100.

$$\frac{\text{mass of oxygen}}{\text{mass of magnesium oxide}} \times 100\% = \frac{16\,\text{g}}{40\,\text{g}} \times 100\% = 40\%$$

So magnesium oxide is 60% magnesium and 40% oxygen.

Questions

1 a) Find the percentage of hydrogen in water, H_2O, with A_r H = 1, O = 16.

b) Find the percentage of copper in copper oxide, CuO, with A_r Cu = 64, O = 16.

Equations and relative formula masses

Whenever we see a balanced symbol equation (see Chapter 6) we can use it to work out the amounts of substances that react together and also to work out the amounts of substances that are produced.

Atoms are never lost, so a balanced equation must have the same number of atoms of each element on each side of the arrow.

For example, carbon burns in oxygen to form carbon dioxide

carbon + oxygen → carbon dioxide

The chemical formula for this is written as:

$$C + O_2 \rightarrow CO_2$$

The relative formula masses are written like this:

$$\underbrace{12 \quad\quad 32}_{44} \quad\quad 44$$

The total mass on each side is the same. The equation therefore is correct.

Summary

A_r is the relative atomic mass of an element. M_r is the relative formula mass of a compound. M_r is found by adding together the A_rs of the elements in the formula. The percentage of an element present in a compound can be found from the formula of the compound. In a balanced symbol equation, the total relative masses on each side of the equation are the same.

Key words

formula This shows the numbers of atoms which combine in a compound.

relative atomic mass, A_r This is the mass of one atom of an element compared with the mass of one atom of hydrogen.

relative formula mass, M_r This is obtained by adding up the relative atomic masses of all the atoms in the formula.

End of Chapter 5 questions

1 The formula of ammonium sulphate is $(NH_4)_2SO_4$.
- a) What four elements are present in this compound?
- b) Say how many atoms of each element combine together.
- c) Use the Periodic Table to find the relative atomic masses of each of the elements in ammonium sulphate.
- d) Work out the relative formula mass of ammonium sulphate.

2 One ore of iron contains iron oxide which has the formula Fe_2O_3.
- a) Use the Periodic Table (page 185) to find the relative atomic masses of iron and oxygen.
- b) How many atoms of iron and oxygen combine together?
- c) What is the relative formula mass of Fe_2O_3?
- d) What is the percentage of iron in this oxide?

3 Zinc reacts with copper oxide according to the equation:

$$\mathbf{Zn + CuO \rightarrow ZnO + Cu}$$

- a) Use the Periodic Table to find the relative atomic masses of zinc, copper and oxygen.
- b) Work out the relative formula masses of copper oxide, CuO, and zinc oxide, ZnO.
- c) Add up the relative masses of the substances on either side of the equation arrow.
- d) What do you notice about the relative masses of the substances on either side of the equation?

6 Chemical change

Chemical reactions happen all around us all the time. **Chemical changes** happen when the atoms in substances are re-arranged to form new substances.

the gas flame gives out heat due to a chemical reaction with oxygen in air
chemical reactions take place every second inside our bodies to keep us alive
cooking food produces many different chemical reactions
chemical reactions break down the food we eat so we can digest it

A chemical reaction changes one substance into one or more new substances. The new substances formed by a chemical reaction will have different properties. After a chemical change it is usually hard to get back to the substance you started with. Cooking an egg is a chemical reaction.

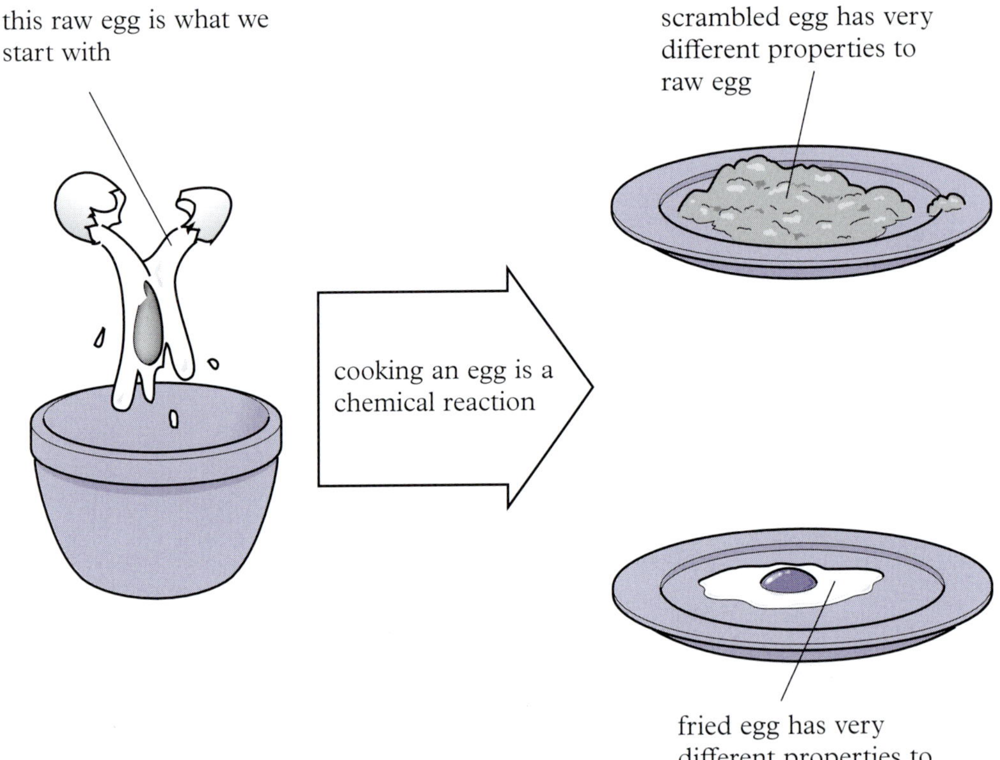
this raw egg is what we start with
scrambled egg has very different properties to raw egg
cooking an egg is a chemical reaction
fried egg has very different properties to raw egg

Physical changes

Turning boiling water which is a liquid into steam which is a gas is not a chemical change because you haven't made a new substance. The water is the same 'stuff' in a different place and in a different state. The three states of matter are gas, liquid and solid. Dissolving salt in water is not a chemical change, because you start with salt and water and you end up with salt and water. The only difference is that the salt and water are mixed together. You haven't made a new substance, even though it looks different. Boiling and dissolving are examples of **physical changes**.

Questions

Which of the following involve a chemical reaction?

a) peeling potatoes,
b) setting off a firework,
c) cooking bacon,
d) sawing wood,
e) making a drink from orange squash,
f) burning branches in a bonfire,
g) separating the colours in ink by chromatography.

Equations

When we have a chemical reaction we often call the original substances the starting materials or '**reactants**', and the substances we end up with the '**products**'.

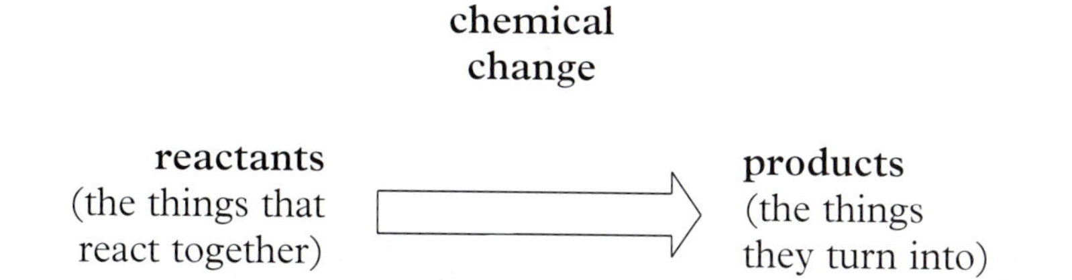
chemical
change
reactants
(the things that
react together)
products
(the things
they turn into)

The arrow in this equation means 'changes into'. A chemical change is a bit like starting with models made out of model bricks, breaking them up and making new models from the same pieces.

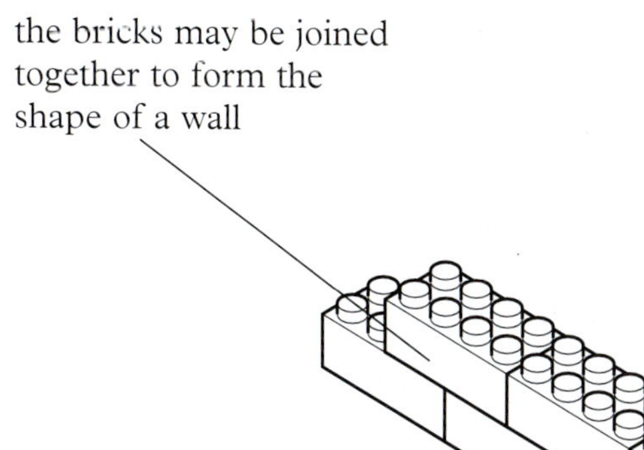
the bricks may be joined
together to form the
shape of a wall

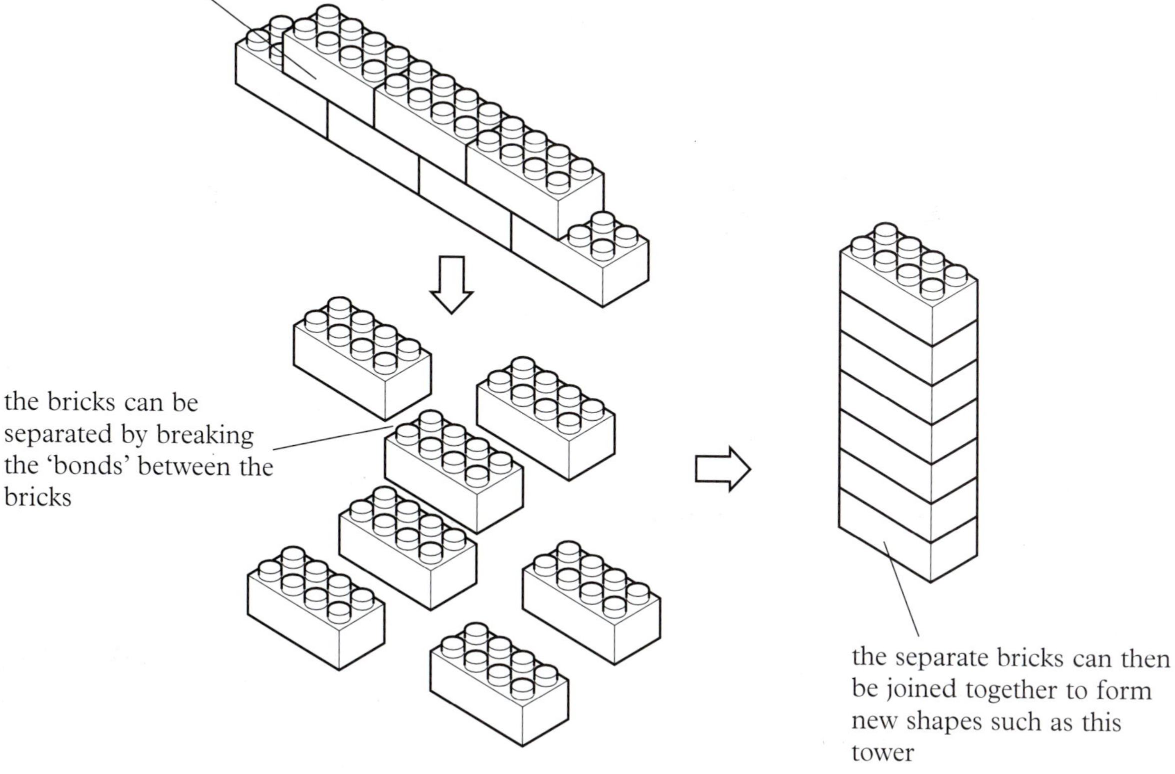
the bricks can be
separated by breaking
the 'bonds' between the
bricks
the separate bricks can then
be joined together to form
new shapes such as this
tower

Although a new shape has been made, no new bricks were used to make the shape and no bricks were lost either.

It is the same with the atoms in a chemical reaction. No atoms are made or lost. They are just put together in a different way.

Once we know that a chemical reaction takes place and we also know what we start with (the reactants) and what we finish up with (the products), then we can write an equation to represent the reaction.

Word equations

A word equation uses chemical names to show the reactants and the products of a chemical reaction. For example, you have probably burnt a strip of magnesium ribbon, which is a silver metal. The magnesium reacts with oxygen which is in the air. This is an example of a chemical change.

the **magnesium ribbon** burns with a very bright light

the magnesium ribbon ends up as a white powder, which is **magnesium oxide**

This chemical change can be represented by a word equation:

magnesium + oxygen ➡ magnesium oxide

The arrow means 'changes into'. The word equation does not tell us how the reaction was done or what happened during the reaction.

There are many different sorts of chemical reactions. Burning magnesium in oxygen is called an **oxidation reaction** because oxygen has been added to the magnesium.

Questions

Write word equations for the following:

a) Burning magnesium ribbon in air to make magnesium oxide.
b) Heating together iron powder and sulphur to make iron sulphide.
c) Starting with copper, heating it in the oxygen in air to make copper oxide.
d) Heating iron in chlorine gas to make iron chloride.
e) Making iron oxide from iron. (Hint: what substance will you need as well as iron?).

Another oxidation reaction happens when you burn charcoal on a barbecue.

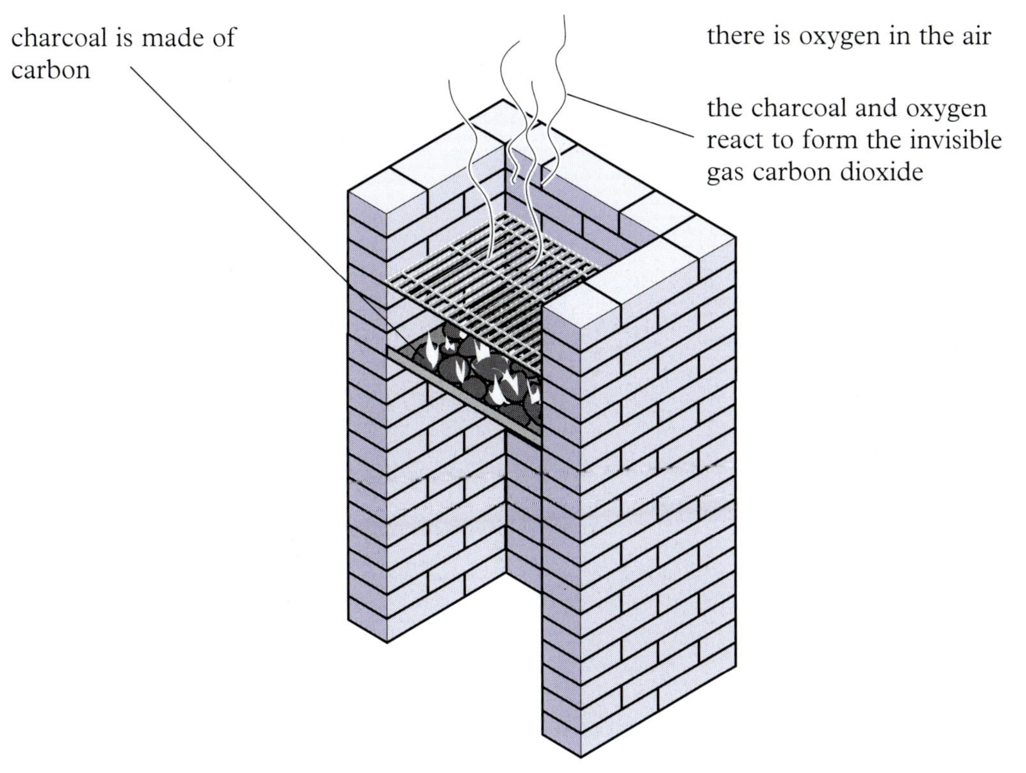
charcoal is made of carbon
there is oxygen in the air
the charcoal and oxygen react to form the invisible gas carbon dioxide

The word equation is:

carbon + oxygen ➡ carbon dioxide

In terms of atoms, we can show the reaction like this:

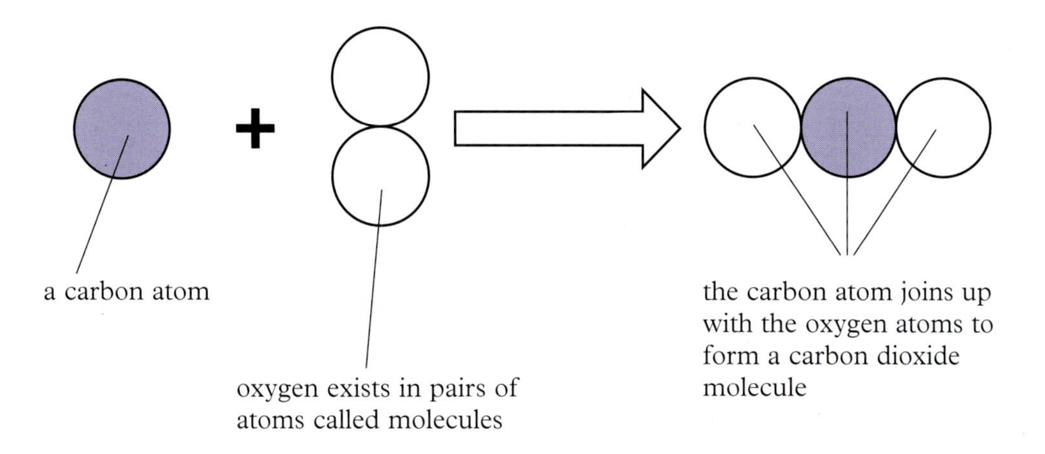
a carbon atom
oxygen exists in pairs of atoms called molecules
the carbon atom joins up with the oxygen atoms to form a carbon dioxide molecule

In this reaction we do not lose or gain any atoms, we just rearrange them.

Symbol equations

Symbol equations are a shorthand method of showing chemical reactions.

The rules for chemical formulae

A **small number after the symbol** tells us how many **atoms of that element** are present. If there is no number there is just one atom.

Water is written as H_2O

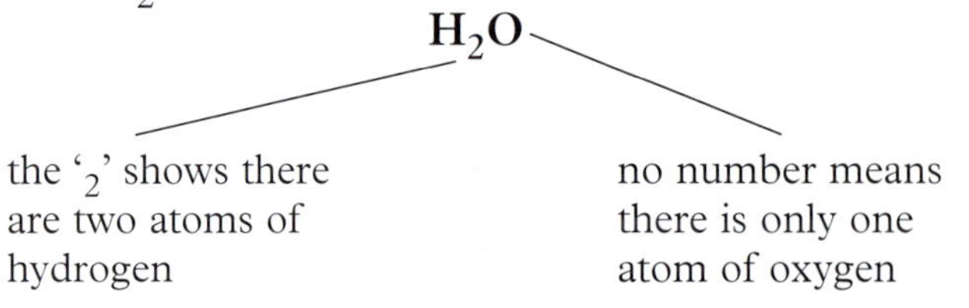
H_2O
the '$_2$' shows there are two atoms of hydrogen
no number means there is only one atom of oxygen

A **small number after a bracket multiplies all the atoms inside the bracket.**

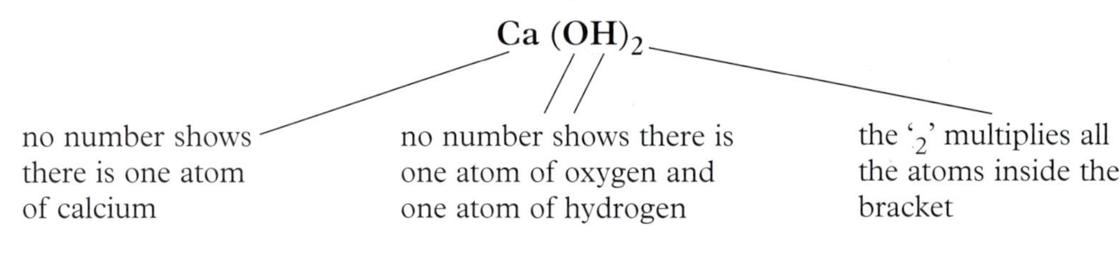
$Ca\,(OH)_2$
no number shows there is one atom of calcium
no number shows there is one atom of oxygen and one atom of hydrogen
the '$_2$' multiplies all the atoms inside the bracket

A **large number in front** of the formulae **multiplies all the atoms in the formula**.

$3H_2O$

the large number three in front of the formula means that H_2 is multiplied by 3 ($3 \times 2 = 6$ H atoms)

the large number three in front of the formula means that the O is also multiplied by 3 ($3 \times 1 = 3$ O atoms)

Using formulae

Carbon has the formula C, oxygen the formula O_2. The '$_2$' means that oxygen atoms go round in pairs. Carbon dioxide has the formula CO_2, which means a group with one carbon atom and two oxygen atoms. We can use these formulae to write the reaction of carbon burning in the oxygen of the air as:

C	+	O_2	➡	CO_2
carbon		**oxygen**		**carbon dioxide**
Reactants				**Product**

This is a shorthand way of saying 'each carbon atom reacts with one molecule of oxygen to produce one molecule of carbon dioxide'.

There are the same number of carbon atoms and oxygen atoms on each side of the equation, so it called a **balanced symbol equation**. The word 'balanced' means that there are the same number of atoms of each element on both sides of the arrow. The word 'symbol' means that we are using chemical symbols rather than words.

State symbols

State symbols are letters in brackets placed after the formulae. These letters give more information about the state of starting materials and the products.

(s) stands for solid – remember that powders are solids.

(l) stands for liquid – like water or oil.
Notice that this has a different meaning from (aq).

(g) stands for gas.

(aq) stands for aqueous, which means 'dissolved in water'.
A solution is often a solid dissolved in water.

We can use state symbols to give more information about a reaction. The burning of charcoal can be shown as below:

$$C(s) + O_2(g) \rightarrow CO_2(g)$$

carbon oxygen carbon dioxide

Other chemical reactions

If you take oxygen away from a compound the reaction is called a **reduction**. This means that reduction is the opposite of oxidation. If we pass hydrogen over hot copper oxide we can reduce the copper oxide to copper.

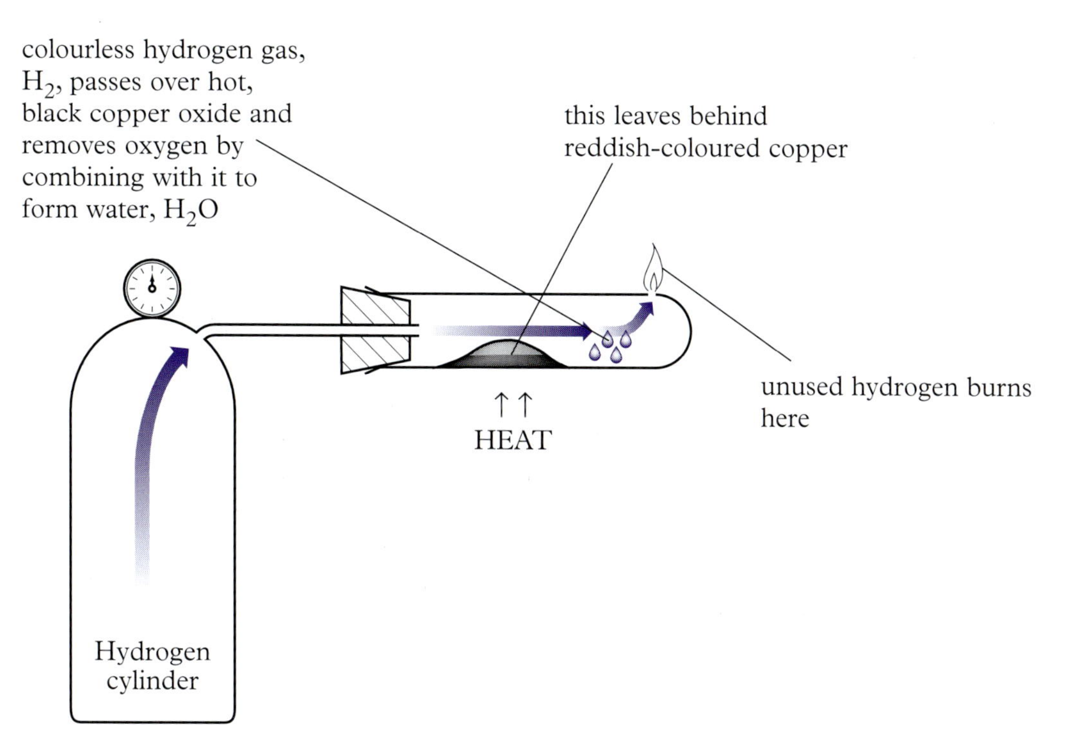
colourless hydrogen gas, H_2, passes over hot, black copper oxide and removes oxygen by combining with it to form water, H_2O
this leaves behind reddish-coloured copper
unused hydrogen burns here
↑↑
HEAT
Hydrogen cylinder

The word equation is:

copper oxide + hydrogen ⟶ copper + water

The symbol equation including state symbols is:

$$CuO(s) + H_2(g) \longrightarrow Cu(s) + H_2O(l)$$

You need to check that the equation is **balanced.** This means there are the same number of atoms of each element on both sides of the equation.

Cu	O	+	H_2	⟶	Cu	+	H_2	O
↑	↑		↑		↑		↑	↑
1 copper	1 oxygen		2 hydrogen		1 copper		2 hydrogen	1 oxygen

By adding up the number of atoms of each element we can see that both sides of the equation have the same number. This equation is therefore balanced.

Displacement reactions

Another kind of reaction is called a **displacement reaction**. If zinc powder is added to copper sulphate solution, the zinc removes or displaces the copper from the copper sulphate.

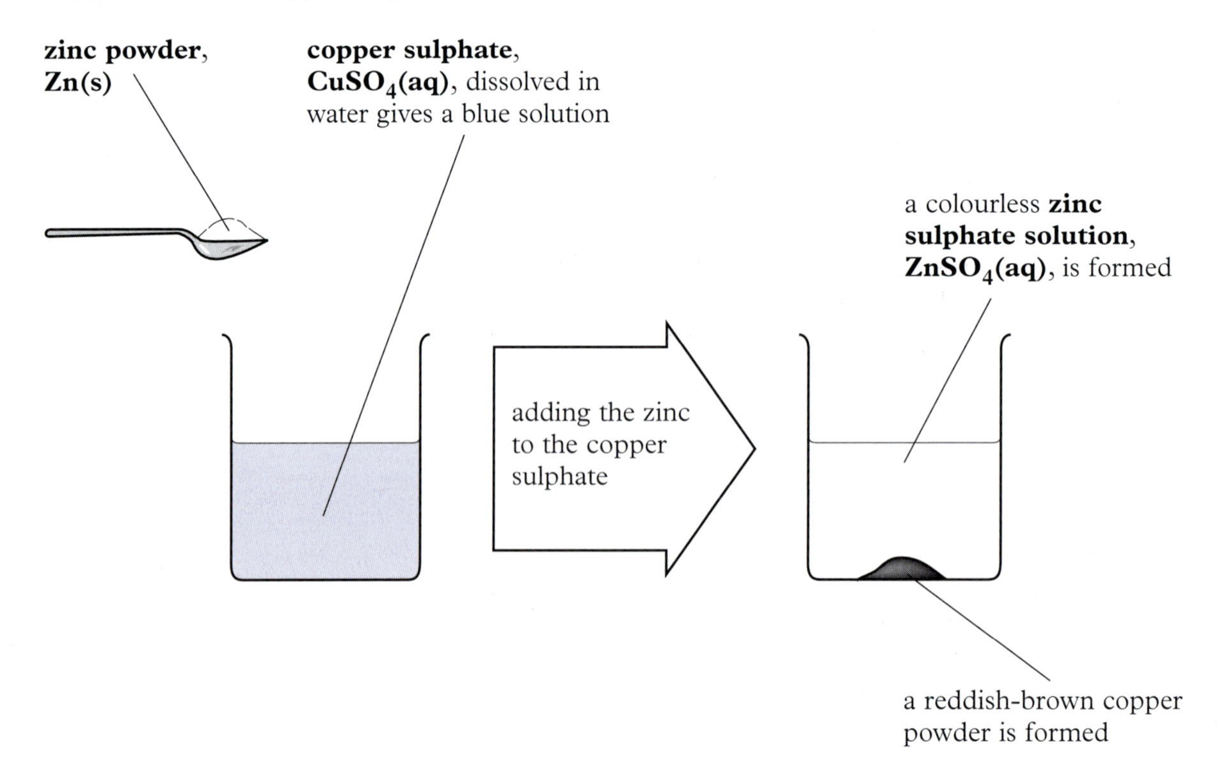
zinc powder, Zn(s)
copper sulphate, CuSO4(aq), dissolved in water gives a blue solution
adding the zinc to the copper sulphate
a colourless zinc sulphate solution, ZnSO4(aq), is formed
a reddish-brown copper powder is formed

The word equation is:

zinc + copper sulphate ➡ zinc sulphate + copper

The symbol equation including state symbols is

$Zn(s)$ +	$CuSO_4(aq)$	➡	$ZnSO_4(aq)$	+ $Cu(s)$
zinc	**copper sulphate**		**zinc sulphate**	**copper**

You need to check that it is balanced. This means there are the same number of atoms of each element on both sides of the equation.

Zn +	Cu	S	O_4	➡	Zn	S	O_4	+ Cu
↑	↑	↑	↑		↑	↑	↑	↑
1 zinc	**1 copper**	**1 sulphur**	**4 oxygen**		**1 zinc**	**1 sulphur**	**4 oxygen**	**1 copper**

This equation is balanced.

Questions

Say which of the following equations are balanced.

a) $H_2 + O_2 \longrightarrow H_2O$
b) $H_2 + Cl_2 \longrightarrow 2HCl$
c) $N_2 + O_2 \longrightarrow NO_2$
d) $NaOH + HCl \longrightarrow NaCl + H_2O$

Summary

A chemical reaction can be written as **reactants ➡ products**.

After a chemical reaction the products will be different from the reactants. The products have different chemical properties from the reactants. Chemical reactions can be written using word equations. Chemical reactions can be written using balanced symbol equations. State symbols tell us whether the reactants and products are solid (s), liquid (l), gas(g) or dissolved in water (aq).

Key words

balanced symbol equation	A symbol equation showing a chemical reaction. There are the same number of atoms of each element on both sides of the equation.
products	The substances that result from a chemical reaction
reactants	The substances that you start with at the beginning of a chemical reaction.

End of Chapter 6 questions

1

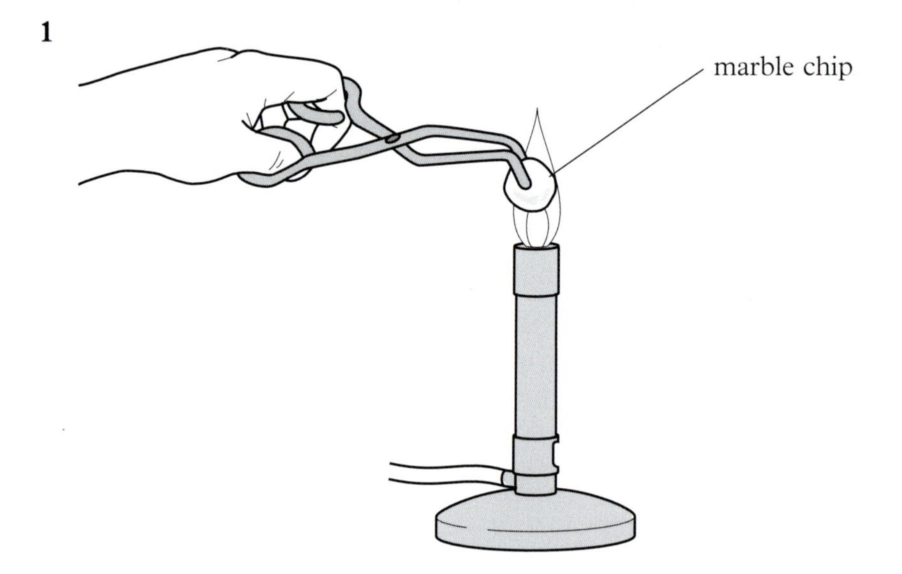
marble chip

When you heat a marble chip it loses the gas carbon dioxide and turns into solid calcium oxide. The chemical name for marble is calcium carbonate. Calcium carbonate is $CaCO_3$, calcium oxide is CaO, carbon dioxide is CO_2.

a) Copy and complete the word equation which shows what happens when the marble is heated.

calcium carbonate → calcium oxide + _________

b) Write the symbol equation with state symbols for the reaction.

c) Is your equation balanced?

d) After heating, will the marble chip become heavier, become lighter or stay the same? Explain your answer.

2.

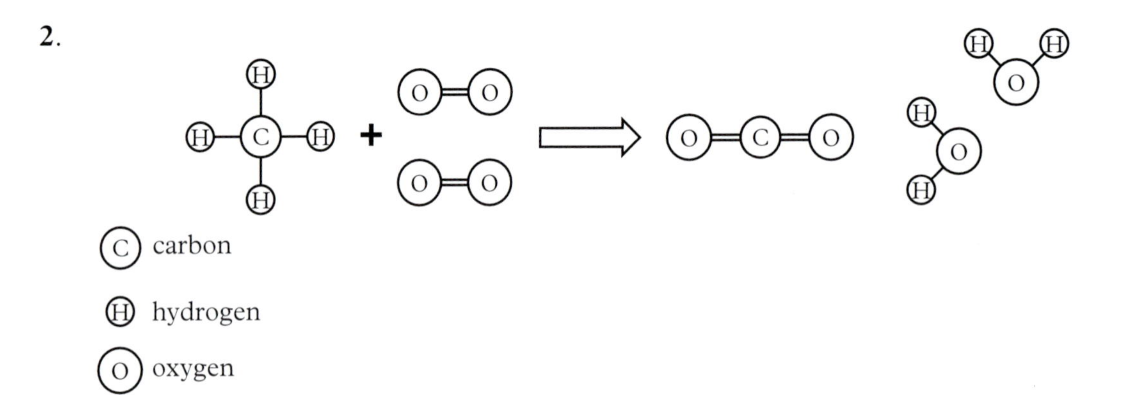
H
H
C
H
H
+
O
O
O
O
O
C
O
H
O
H
H
O
H
C carbon
H hydrogen
O oxygen

The diagram above shows how methane burns in oxygen in the air to make water and carbon dioxide.

a) Write a word equation for this reaction.

b) Use the diagram to write a balanced symbol equation. Hint: notice how many molecules of oxygen are needed.

3 Look carefully at the following equation:

$$CaCO_3(s) + HCl(aq) \rightarrow CaCl_2(aq) + H_2O(l) + CO_2(g)$$

a) What do the following symbols in the equation above mean?
   i) (s)
   ii) (l)
   iii) (g)
   iv) (aq)

b) Is the equation balanced? Explain your answer.

c) $CaCO_3$ is the formula for calcium carbonate.
   i) What is H_2O the formula for?
   ii) What is CO_2 the formula for?

d) Copy and complete the word equation below.

calcium carbonate + hydrochloric acid → calcium chloride + ________ + ________

4 Look at the reactions below.

magnesium + oxygen → magnesium oxide

copper oxide + hydrogen → copper + water

copper sulphate + zinc → copper + zinc sulphate

methane + oxygen → carbon dioxide + water

For each reaction, choose the word which best describes it from the following list. You may use each word once, more than once or not at all.

neutralisation reduction oxidation

mixing displacement

7 Acids and alkalis

Acids and alkalis are all around us. Vinegar and fruit juices are acids, while oven cleaners are alkalis. We all know the sharp or sour taste of a lemon. This is the typical acid taste. Alkalis cancel out the acidity of acids. Alkalis feel soapy. However we do not test for acids and alkalis by tasting or feeling them because they may be dangerous.

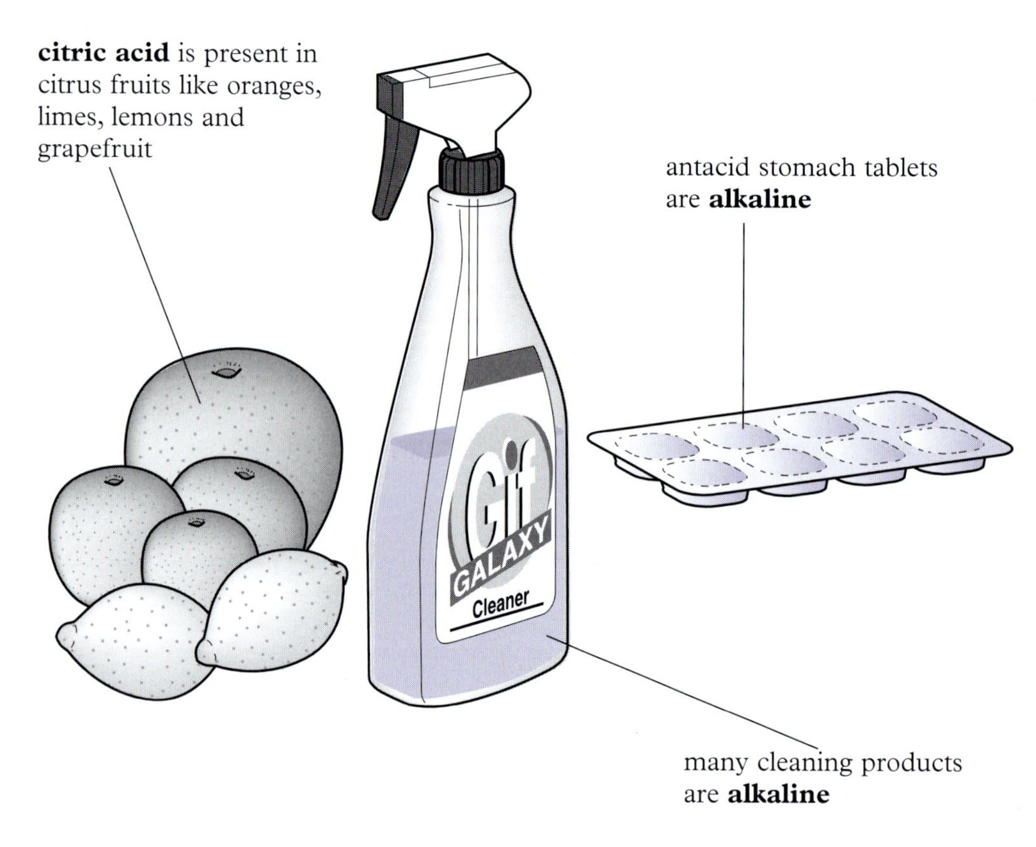
citric acid is present in citrus fruits like oranges, limes, lemons and grapefruit
antacid stomach tablets are **alkaline**
Cif
GALAXY
Cleaner
many cleaning products are **alkaline**

Indicators

We use **indicators** to test for acids and alkalis. Indicators are either solutions or pieces of paper that change colour when they are added to acids or alkalis.

There are different sorts of indicators.

Indicator	Colour in acid solution	Colour in neutral solution	Colour in alkaline solution
Litmus	Red	Purple	Blue
Phenolphthalein	Colourless	Pink	Bright pink
Methyl orange	Red	Orange	Yellow

A useful indicator is called **Universal Indicator**. It has a range of colours and can tell us how acidic or alkaline a solution is.

Testing with Universal Indicator

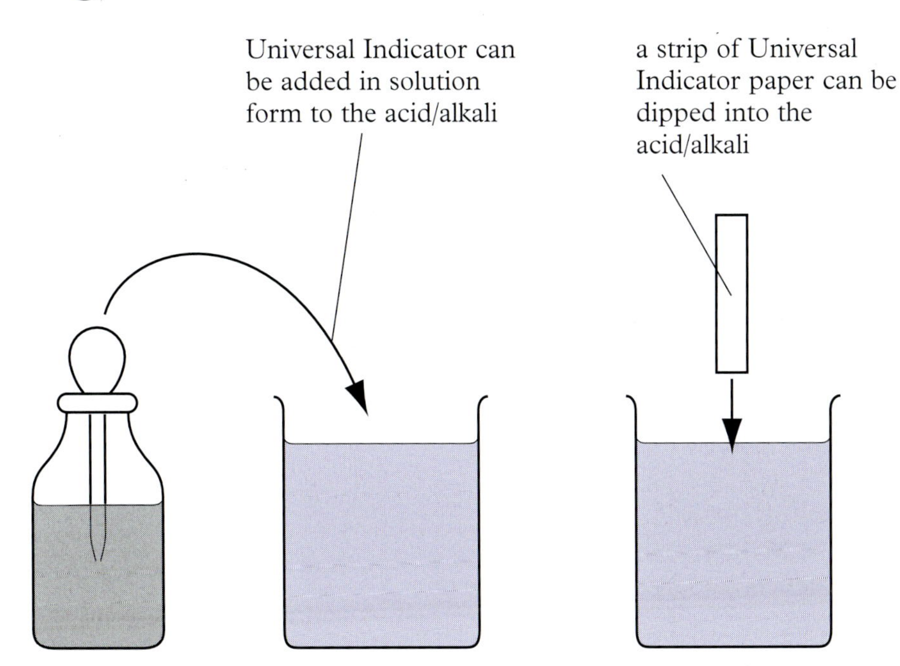
Universal Indicator can be added in solution form to the acid/alkali
a strip of Universal Indicator paper can be dipped into the acid/alkali

Universal Indicator solution and paper change colour when they come in contact with an acid or an alkali. The colour they change to depends on how acid or alkaline the solution is.

The pH scale

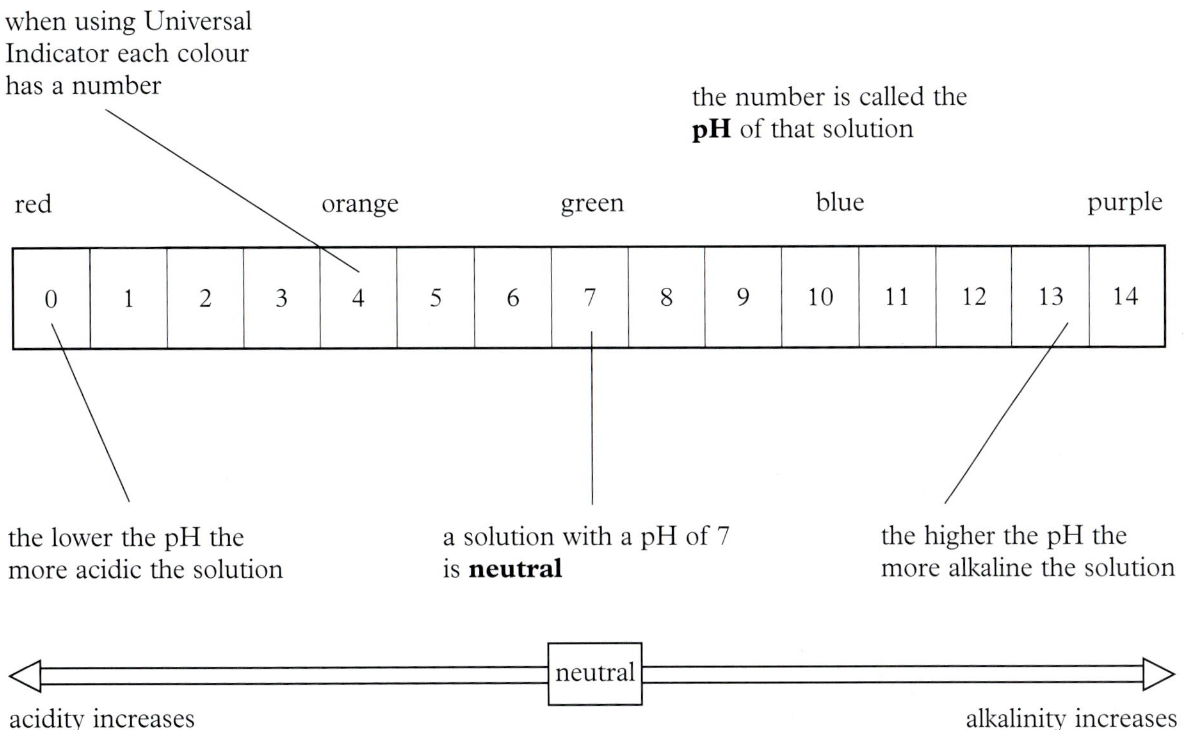
when using Universal Indicator each colour has a number
the number is called the **pH** of that solution
red
orange
green
blue
purple
0
1
2
3
4
5
6
7
8
9
10
11
12
13
14
the lower the pH the more acidic the solution
a solution with a pH of 7 is **neutral**
the higher the pH the more alkaline the solution
neutral
acidity increases
alkalinity increases

The reactions of dilute acids

The acids we find in the science laboratory can be used to make salts.

Name of the acid	Name of the salt it forms
Sulphuric acid	**sulph**ate
Hydro**chlor**ic acid	**chlor**ide
Nitric acid	**nitr**ate

Concentrated acids and alkalis are very dangerous. They both cause blistering and burns on the skin. The dilute acids we use in the laboratory are all solutions in water. Some acids are more concentrated than others.

Acids form salts when they react with:
metals e.g. magnesium
metal oxides e.g. magnesium oxide
metal hydroxides e.g. magnesium hydroxide
metal carbonates e.g. magnesium carbonate

Reactions of acids with metals

Dilute acids react with reactive metals like magnesium or zinc to form **salts**. When the reaction takes place hydrogen gas is given off. This reaction may be written as an equation:

Example

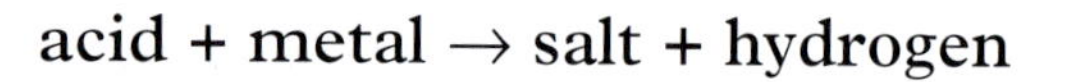

acid + metal → salt + hydrogen

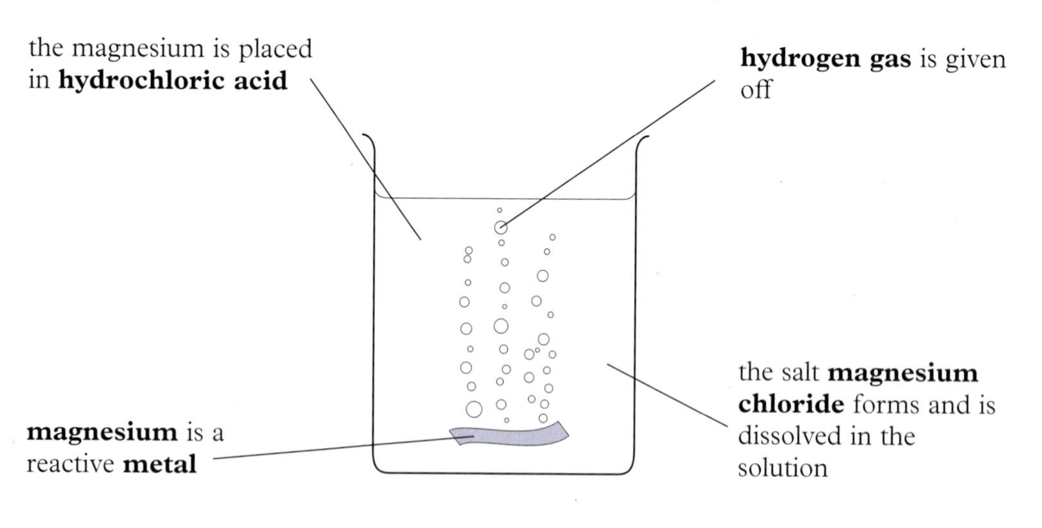
the magnesium is placed in hydrochloric acid
hydrogen gas is given off
the salt magnesium chloride forms and is dissolved in the solution
magnesium is a reactive metal

The word equation for this reaction is:

hydrochloric acid + magnesium → magnesium chloride + hydrogen

acid + metal → salt + hydrogen

Some metals are known as **unreactive** because they do not react with dilute acids. Copper and silver are both unreactive metals. Metals which do react with dilute acids are called **reactive**.

Reactions of acids with metal oxides

Magnesium oxide is formed when magnesium reacts with oxygen. Magnesium oxide is a metal oxide. Acids react with metal oxides to form **salts** and **water**.

acid + metal oxide → salt + water

Example

sulphuric acid + magnesium oxide → magnesium sulphate + water

acid + metal oxide → salt + water

Reactions of acids with metal hydroxides

Magnesium hydroxide is a metal hydroxide. Metal hydroxides contain metal, hydrogen and oxygen. Acids react with metal hydroxides to form salts and water.

acid + metal hydroxide → salt + water

Example

sulphuric acid + magnesium hydroxide → magnesium sulphate + water

acid + metal hydroxide → salt + water

Questions

1 a) What colour does litmus paper change to when it is dipped into an acidic solution?
b) What colour does litmus paper change to when it is dipped into an alkaline solution?
c) If a solution has a pH of 2 is it acidic, neutral or alkaline?

2 a) What is the name of the salt which sulphuric acid forms?
b) Write a word equation for the reaction of sulphuric acid with magnesium.

3 Write the word equation for the reaction between magnesium oxide and hydrochloric acid.

4 Write the word equation for the reaction between sodium hydroxide and hydrochloric acid.

Reactions of acids with metal carbonates

Magnesium carbonate is a metal carbonate. Metal carbonates contain metal, carbon and oxygen. Acids react with metal carbonates to form salts, water and carbon dioxide.

acid + metal carbonate → salt + water + carbon dioxide

Example
Magnesium carbonate reacts with hydrochloric acid to form magnesium chloride, water and carbon dioxide gas.

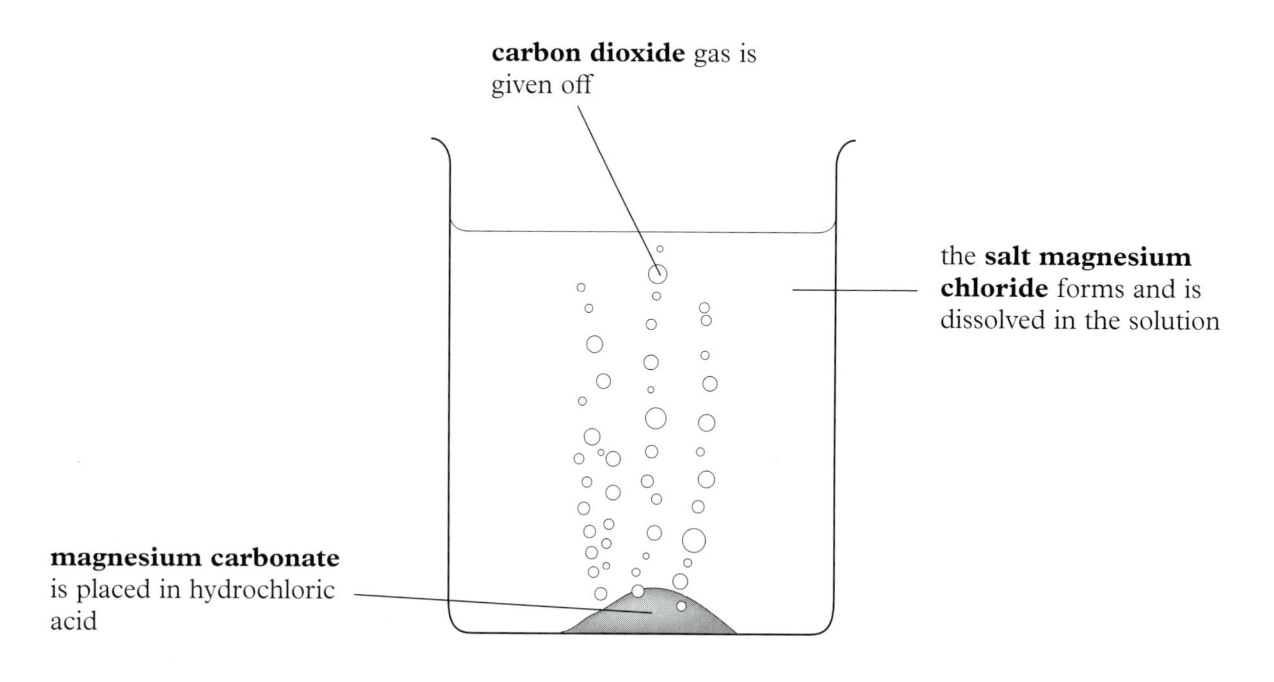
carbon dioxide gas is given off
the **salt magnesium chloride** forms and is dissolved in the solution
magnesium carbonate is placed in hydrochloric acid

The word equation for this reaction is:

hydrochloric acid	**+**	**magnesium carbonate**	**→**	**magnesium chloride**	**+**	**water**	**+**	**carbon dioxide**
acid	+	metal carbonate	→	salt	+	water	+	carbon dioxide

Questions

1 Copy and complete this table:

Metal	Acid	Gas given off? Yes or No	Name of gas
zinc	hydrochloric		
copper oxide	hydrochloric		
copper carbonate	sulphuric		
magnesium hydroxide	sulphuric		

Neutralisation

When an acid reacts with an alkali, we call the reaction a neutralisation. In the laboratory, we can find how much acid will neutralise an alkali by performing a **titration**, as shown in the diagram.

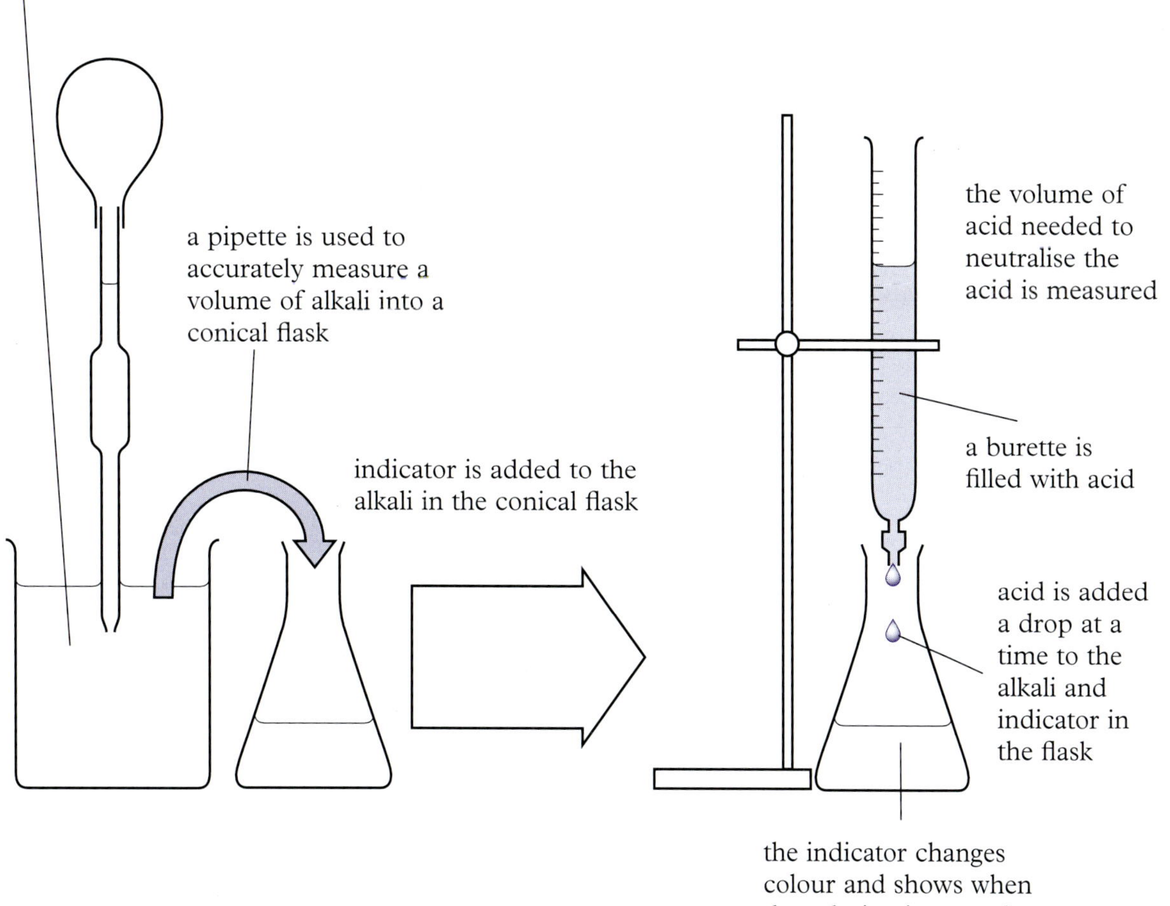
we start with a sample of alkali
a pipette is used to accurately measure a volume of alkali into a conical flask
indicator is added to the alkali in the conical flask
the volume of acid needed to neutralise the acid is measured
a burette is filled with acid
acid is added a drop at a time to the alkali and indicator in the flask
the indicator changes colour and shows when the solution is neutral

The more acid you need to add to reach the neutral point, the more alkaline was the original solution.

Useful examples of neutralisation

Indigestion tablets are taken to get rid of acid indigestion. They contain alkaline chemicals like sodium hydrogencarbonate (which is often called sodium bicarbonate) or magnesium hydroxide which will neutralise any excess hydrochloric acid that is made in our stomachs.

Farmers spread lime, which is calcium oxide, onto acidic soils to neutralise the acidity.

Questions

1 Copy the sentences below, filling in the missing words from the following list:

conical colour burette acid neutralise

You could perform a titration to compare the alkalinity of indigestion tablets.

a) You would put a ground-up indigestion tablet into the c______ flask, with a little water and a few drops of indicator.

b) Then you would run acid from the b______ into the flask until the indicator changes c______.

c) You would note the amount of acid that is needed to n______ the tablet. Then you would repeat the whole experiment with a different type of indigestion tablet. The tablet that needed the most a______ to neutralise it must be the most alkaline.

Sulphuric acid – an important acid

Sulphuric acid is used in many different industries. This pie-chart shows which industries sulphuric acid is used in.

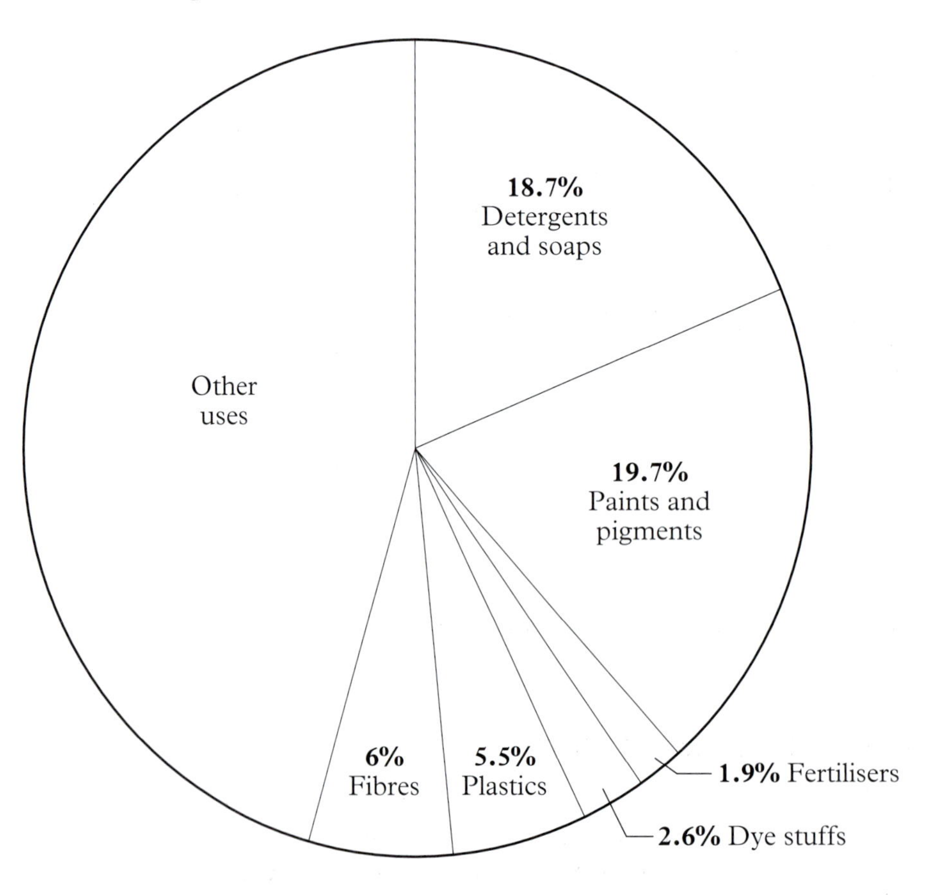
18.7%
Detergents and soaps
19.7%
Paints and pigments
1.9% Fertilisers
2.6% Dye stuffs
5.5%
Plastics
6%
Fibres
Other uses

Sulphuric acid is made by the **Contact Process**.

The raw materials are sulphur, air and water. All of these are cheap or free. The Contact Process turns the raw materials into sulphuric acid.

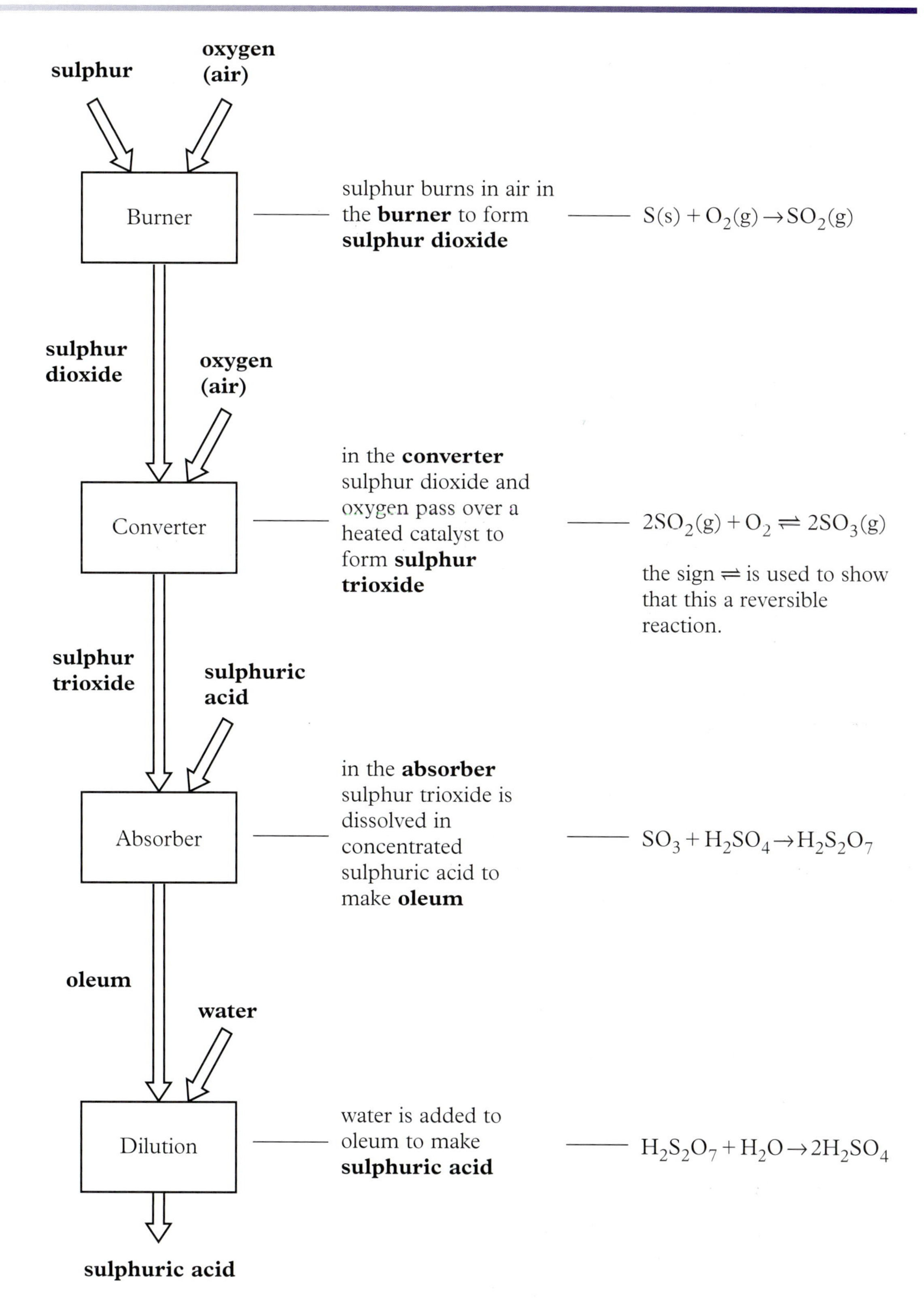
sulphur
oxygen (air)
Burner
sulphur burns in air in the **burner** to form **sulphur dioxide**
$S(s) + O_2(g) \rightarrow SO_2(g)$
sulphur dioxide
oxygen (air)
Converter
in the **converter** sulphur dioxide and oxygen pass over a heated catalyst to form **sulphur trioxide**
$2SO_2(g) + O_2 \rightleftharpoons 2SO_3(g)$
the sign $\rightleftharpoons$ is used to show that this a reversible reaction.
sulphur trioxide
sulphuric acid
Absorber
in the **absorber** sulphur trioxide is dissolved in concentrated sulphuric acid to make **oleum**
$SO_3 + H_2SO_4 \rightarrow H_2S_2O_7$
oleum
water
Dilution
water is added to oleum to make **sulphuric acid**
$H_2S_2O_7 + H_2O \rightarrow 2H_2SO_4$
sulphuric acid

As soon as sulphur trioxide forms, some of it breaks down into sulphur dioxide and oxygen again. This means only some of the sulphur dioxide is converted to sulphur trioxide, and this produces a mixture of all three gases. The temperature and pressure are chosen so that as much sulphur trioxide as possible is made.

A low temperature and high pressure gives the best yield of sulphur trioxide. At low temperatures however, the reaction is slow and high pressures make the production more expensive. This is because the containers have to be strong and it takes energy to keep the pressure high.

Most sulphuric acid plants run at about 450 °C and use ordinary atmospheric pressure. A catalyst called vanadium(V) oxide speeds up the reaction.

Summary

Acids and **alkalis** are all around us. We can test for acids and alkalis by using **indicators**. A very useful indicator is called **Universal Indicator**. It changes colour when it comes into contact with an acid or alkali and the colour it changes to depends on how acidic or alkaline a substance is. The colour corresponds to a number on the **pH scale**. The pH scale shows how acidic or alkaline a substance is. A substance with a **pH of 0 is very acidic**. A substance with a **pH of 7 is neutral**, which means it is neither acidic or alkali. A substance with a **pH of 14 is very alkaline**.

Solutions of acids can react to form **salts**. When an **acid** and a **metal** react, **salt** and **hydrogen** are formed. When an **acid** and a **metal oxide** react, **salt** and **water** are formed. When an **acid** and a **metal hydroxide** react, **salt** and **water** are formed. When an **acid** and a **metal carbonate** react, **salt** and **water** and **carbon dioxide** are formed.

When an acid reacts with an alkali, we call the reaction a **neutralisation**. We can find out how much acid will neutralise an alkali by performing a **titration**.

Sulphuric acid is used in many different industries. Sulphuric acid is made by the **Contact Process**. The raw materials are **sulphur**, **air** and **water**. The Contact Process turns the raw materials into sulphuric acid.

Key words

acids Solutions with a pH of less than 7.

alkalis Solutions with a pH of greater than 7.

Contact Process A series of chemical reactions that produce sulphuric acid.

indicators Dyes that change colour in solutions of different pH.

neutralisation The reaction of an acid and an alkali.

pH scale This scale goes from 0 to 14 and tells us how acidic or alkali a solution is. A pH of 0 is very acidic, a pH of 7 is neutral and a pH of 14 is very alkaline.

titration A method to see how much acid will neutralise an alkali.

Universal Indicator A very useful indicator which changes to a range of colours depending on how acidic or alkaline a substance is. The colour corresponds to a number on the pH scale.

End of Chapter 7 questions

1 a) Fill in the blanks, using words from the following list:

metal carbon dioxide hydrochloric sulphates hydrogen nitric

When acids are neutralised they form salts. Sulphuric acid forms s__________ . H__________ acid forms chlorides. Nitrates are formed from n__________ acid. Reactive metals form salts when they react with acids. The gas h__________ is given off when reactive metals are placed in acid. Metal carbonates react with acids to form salts and water. The gas c__________ is also given off. M__________ oxides and hydroxides react with acids to form salts and water only.

b) Explain how you would you test for hydrogen (Hint: see page 184).
c) Explain how you would test for carbon dioxide (Hint: see page 184).

2 a) Copy and complete the table about the acidity and alkalinity of substances that you might find at home

Substance	pH number	Acidic, alkaline or neutral
Indigestion tablets	9	
Oven cleaner	14	
Lemon juice	5	
Water		neutral

3 Look at the substances listed in the table of question 2.

a) Which one would be dangerous if it spilt onto your skin?
b) What should you do if some were to spill onto your skin?

4 Describe carefully how you would find the pH of a substance.

5 Indigestion tablets are used to neutralise excess acid in the stomach.
a) What is the chemical name for a substance that will neutralise an acid?
b) One brand of tablet says that the active ingredient is calcium carbonate.
i) Write a word equation for the reaction of calcium carbonate with hydrochloric acid which is the acid in stomachs.
ii) One of the products of the reaction of calcium carbonate with hydrochloric acid is a gas. What is the gas?
c) Brand X indigestion tablet was compared with Brand Y indigestion tablet by putting a crushed tablet of each into separate flasks with a little water and adding a few drops of an indicator. Then acid was added from a burette until the indicator changed colour. The results are shown in the table below.

	Brand X	Brand Y
Volume of acid added	$20\,cm^3$	$25\,cm^3$

i) Which tablet contains more alkali? Explain your answer.
ii) Why was each tablet crushed up?
iii) What is an indicator?
iv) Brand X had coloured tablets. What problems might this cause?

6 Oven cleaners often contain the alkali sodium hydroxide, which helps them to remove grease. The diagram below shows an apparatus used to compare the amount of alkali in different brands of oven cleaner.

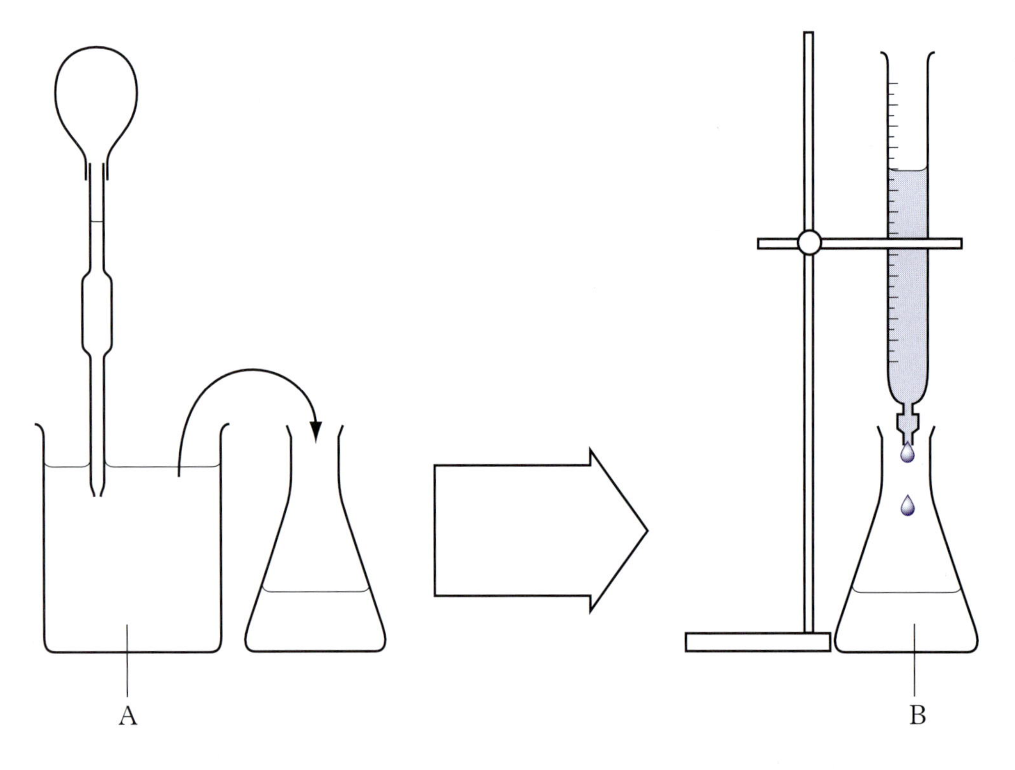
A
B

A sample of oven cleaner is placed in the flask with phenolphthalein indicator and hydrochloric acid is added until the indicator changed colour.

a) Name the pieces of apparatus labelled A and B.
b) What type of reaction is happening between the acid and the alkali?
c) Write a word equation for the reaction between the sodium hydroxide and the hydrochloric acid.
d) Phenolphthalein indicator changes colour from red in alkali to colourless in acid. Explain why this would be a better indicator to use than Universal Indicator.
e) With 10 g of 'Ovenkleen' in the flask, the indicator changed colour after 20 cm^3 of an acid had been added. With 10 g of 'Grimebuster', 25 cm^3 of an acid was needed.
   (i) Name two factors which would have been kept the same for this experiment to be a fair comparison between the two cleaners.
   (ii) If these factors were kept the same, which cleaner contains most sodium hydroxide?
f) Give two sensible safety precautions that people should take when using any cleaner containing sodium hydroxide.

8 Changes to the atmosphere

The gases surrounding the Earth are called the **atmosphere**.

The bottom layer in which we live is made up of a mixture of gases. About 80% of air is made from nitrogen, and about 20% of air is made from oxygen.

Air also contains very small amounts of carbon dioxide, water vapour and **inert gases**.

This mixture of gases in the air is the same now as it was 200 million years ago. This is a long time to us, but the Earth is much older than that. Earth is about 4600 million years old and it was many millions of years before the atmosphere contained the mixture of gases we have today.

How the atmosphere began

The Earth started out as a ball of hot gases. Eventually the hot gases cooled to liquid rock. Over a long period of time the surface cooled to form a thin, solid crust.

4500 million years ago

4500 million years ago there was **no oxygen** in the air. The atmosphere was mainly **carbon dioxide** with some water vapour, nitrogen, methane and ammonia.

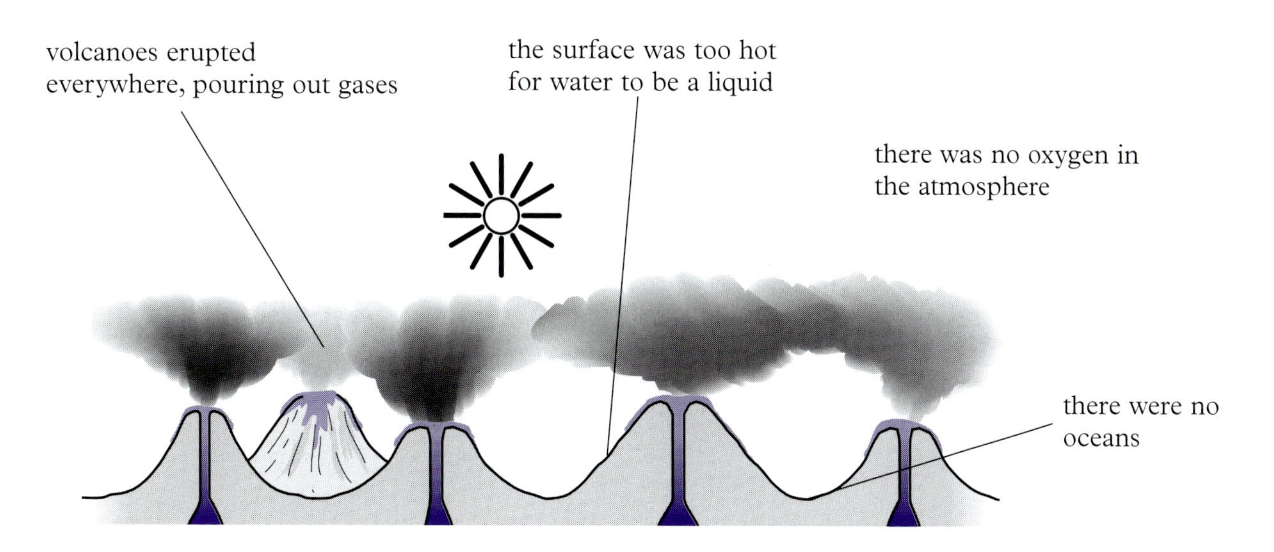
volcanoes erupted everywhere, pouring out gases
the surface was too hot for water to be a liquid
there was no oxygen in the atmosphere
there were no oceans

3000 million years ago

3000 million years ago the surface became **cool** enough for **water** to form as a liquid.

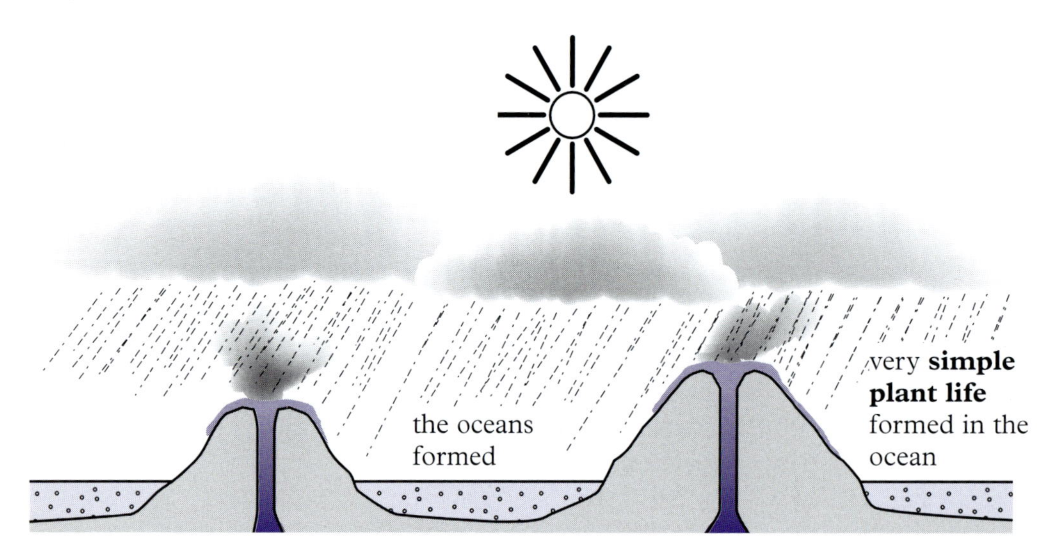
the oceans formed
very **simple plant life** formed in the ocean

2000 million years ago

2000 million years ago oxygen was being released into the atmosphere.

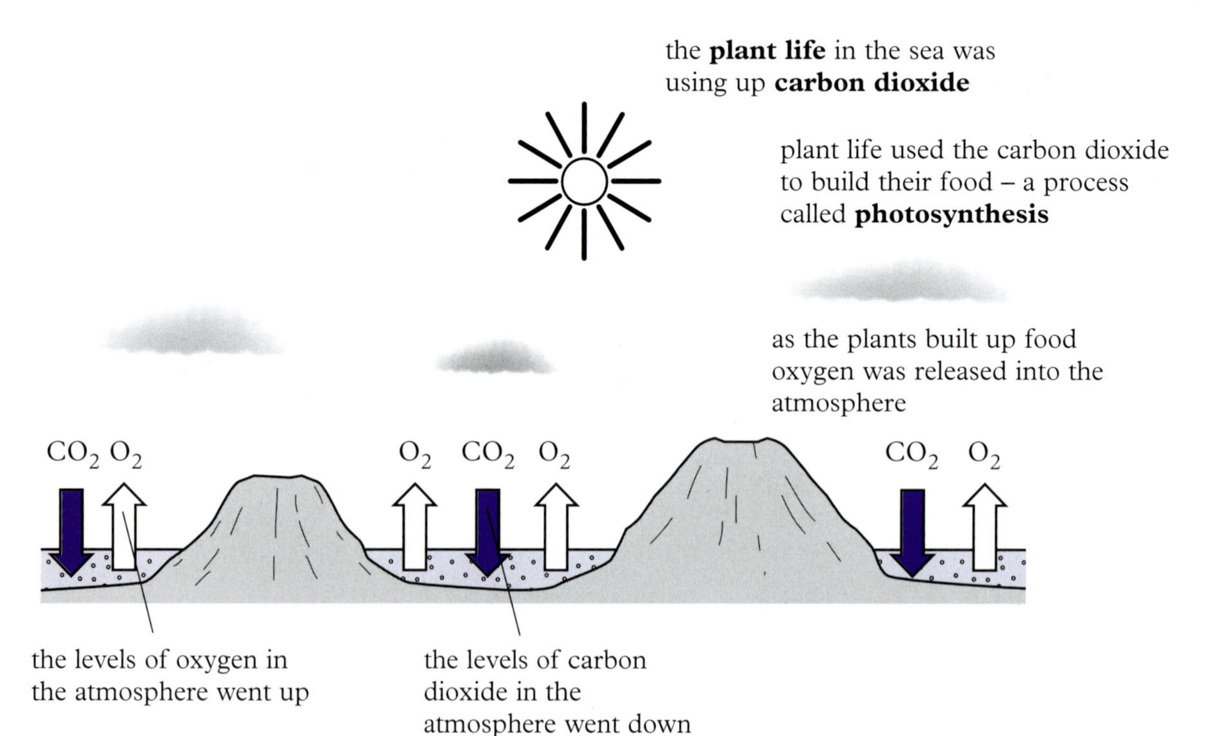
the **plant life** in the sea was using up **carbon dioxide**
plant life used the carbon dioxide to build their food – a process called **photosynthesis**
as the plants built up food oxygen was released into the atmosphere
CO_2 O_2
O_2 CO_2 O_2
CO_2 O_2
the levels of oxygen in the atmosphere went up
the levels of carbon dioxide in the atmosphere went down

1000 million years ago

1000 million years ago **tiny animals** appeared in the sea. The animals used oxygen to breathe. At this time a layer of ozone formed. The ozone helped life to develop.

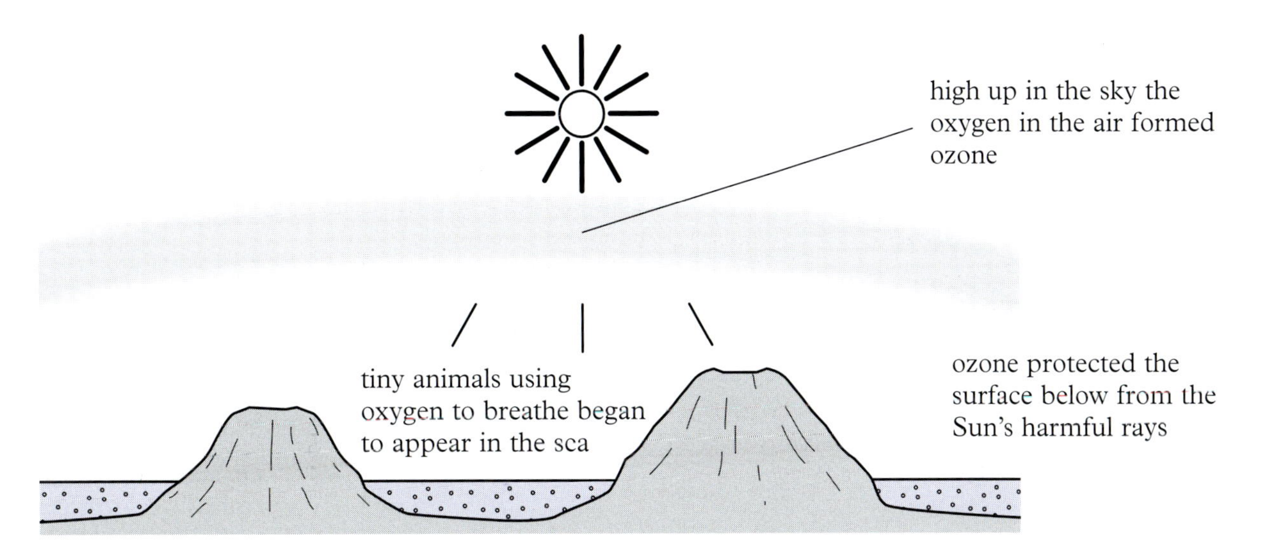
high up in the sky the oxygen in the air formed ozone
tiny animals using oxygen to breathe began to appear in the sea
ozone protected the surface below from the Sun's harmful rays

400 million years ago

400 million years ago the oxygen levels continued to rise. The carbon dioxide levels continued to fall.

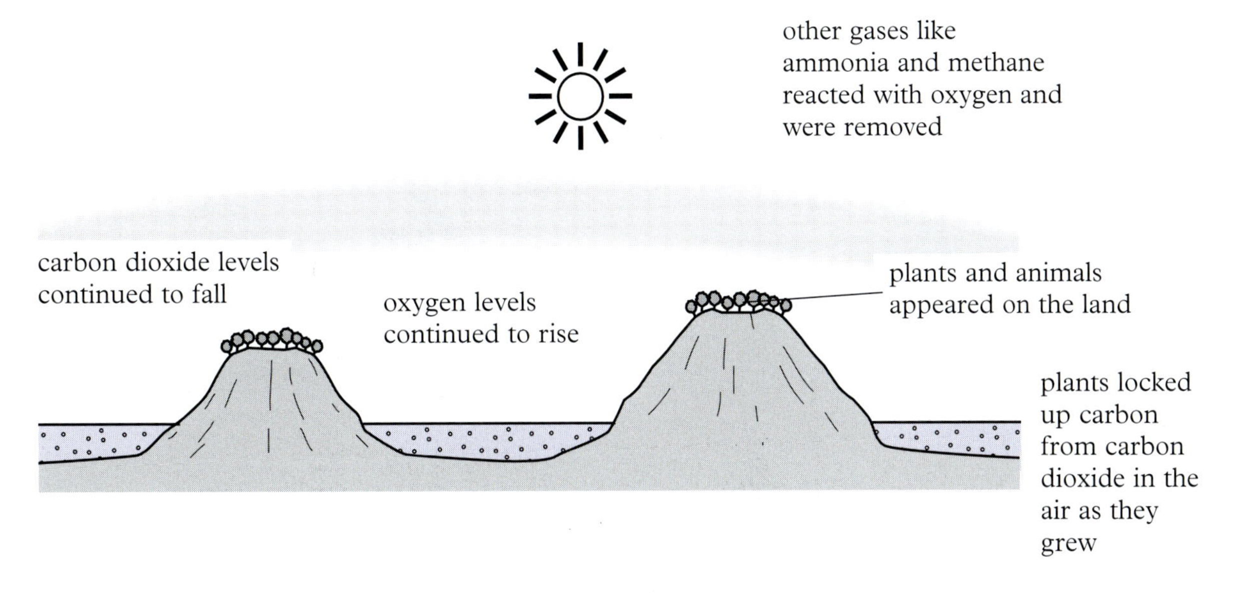
other gases like ammonia and methane reacted with oxygen and were removed
carbon dioxide levels continued to fall
oxygen levels continued to rise
plants and animals appeared on the land
plants locked up carbon from carbon dioxide in the air as they grew

Today

Today, the amounts of gases in the atmosphere stay roughly the same. But all the time the gases are being removed and then replaced.

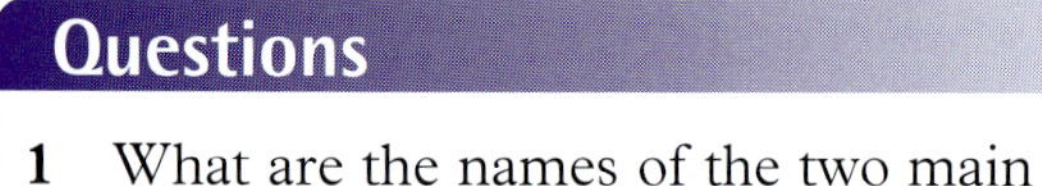

Questions

1 What are the names of the two main gases in our atmosphere today?

2 Why did oceans form once the surface of the Earth cooled?

3 There was no oxygen in the atmosphere when the Earth was formed. Why did the oxygen level start to go up and the carbon dioxide level fall around 2000 million years ago?

4 Why is the ozone layer important?

The water cycle

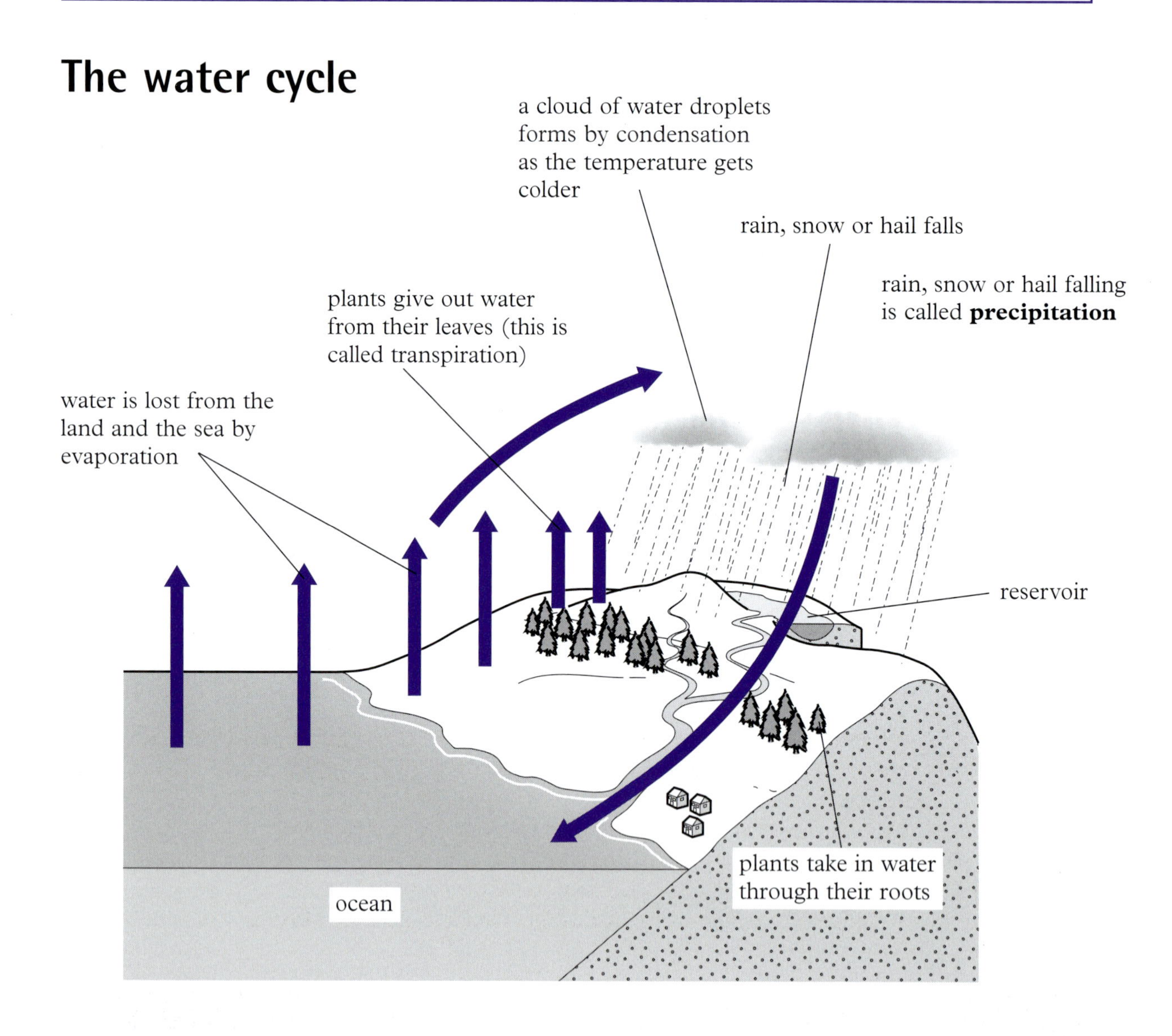
a cloud of water droplets forms by condensation as the temperature gets colder
rain, snow or hail falls
rain, snow or hail falling is called **precipitation**
plants give out water from their leaves (this is called transpiration)
water is lost from the land and the sea by evaporation
reservoir
plants take in water through their roots
ocean

Questions

1 Which process removes water from the Earth's surface into the air?

2 How does water pass from the air to the surface of the Earth?

3 Houses get their water from the reservoir. Explain how water from the houses ends up back in the reservoir.

The nitrogen cycle

Plants need nitrogen to build proteins. Although 80% of the air is nitrogen most plants can't use nitrogen gas directly. Plants can only use nitrogen compounds dissolved in water. Plants absorb these compounds through their roots. Nitrogen doesn't easily dissolve in water and it doesn't react very well with other elements. Some soluble nitrogen compounds are made by chemical reactions during thunderstorms.

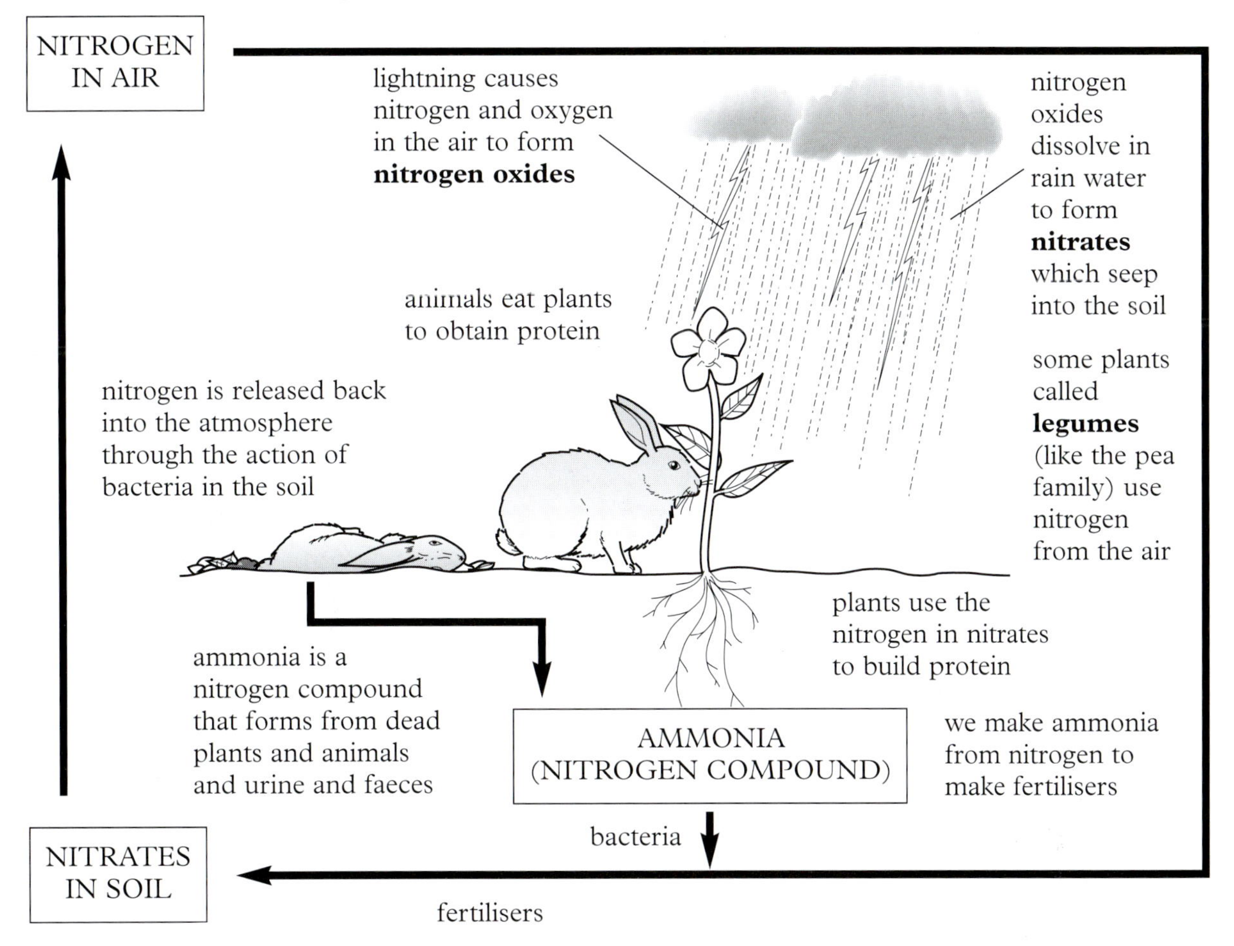
NITROGEN IN AIR
lightning causes nitrogen and oxygen in the air to form **nitrogen oxides**
nitrogen oxides dissolve in rain water to form **nitrates** which seep into the soil
animals eat plants to obtain protein
some plants called **legumes** (like the pea family) use nitrogen from the air
nitrogen is released back into the atmosphere through the action of bacteria in the soil
plants use the nitrogen in nitrates to build protein
ammonia is a nitrogen compound that forms from dead plants and animals and urine and faeces
AMMONIA (NITROGEN COMPOUND)
we make ammonia from nitrogen to make fertilisers
bacteria
NITRATES IN SOIL
fertilisers

Some plants, called legumes, can use nitrogen from the air. The pea plant is a legume. Legumes have bacteria in their roots. The bacteria change nitrogen gas into compounds which can be used by the plant. Humans use an artificial process to convert nitrogen into ammonia. This is called the **Haber Process**.

Questions

1 Why do plants need nitrogen?

2 What does lightning do?

3 How do plants get nitrogen?

4 a) Where does the rabbit in the diagram above obtain protein from?
b) How is ammonia formed in the diagram above?
c) How is nitrogen released back into the atmosphere?

The carbon cycle

All life on Earth is based on carbon. Living things lock up carbon in their cells. Living things act as a store, keeping carbon out of the atmosphere. The stored carbon is released into the atmosphere when the plant or animal dies. In some cases it may stay locked up in the Earth as a fossil fuel.

The fossil fuels we use today are formed from the decay of plants and animals that died millions of years ago. When we burn fossil fuels like coal and crude oil, we release the carbon as carbon dioxide. This is part of the carbon cycle:

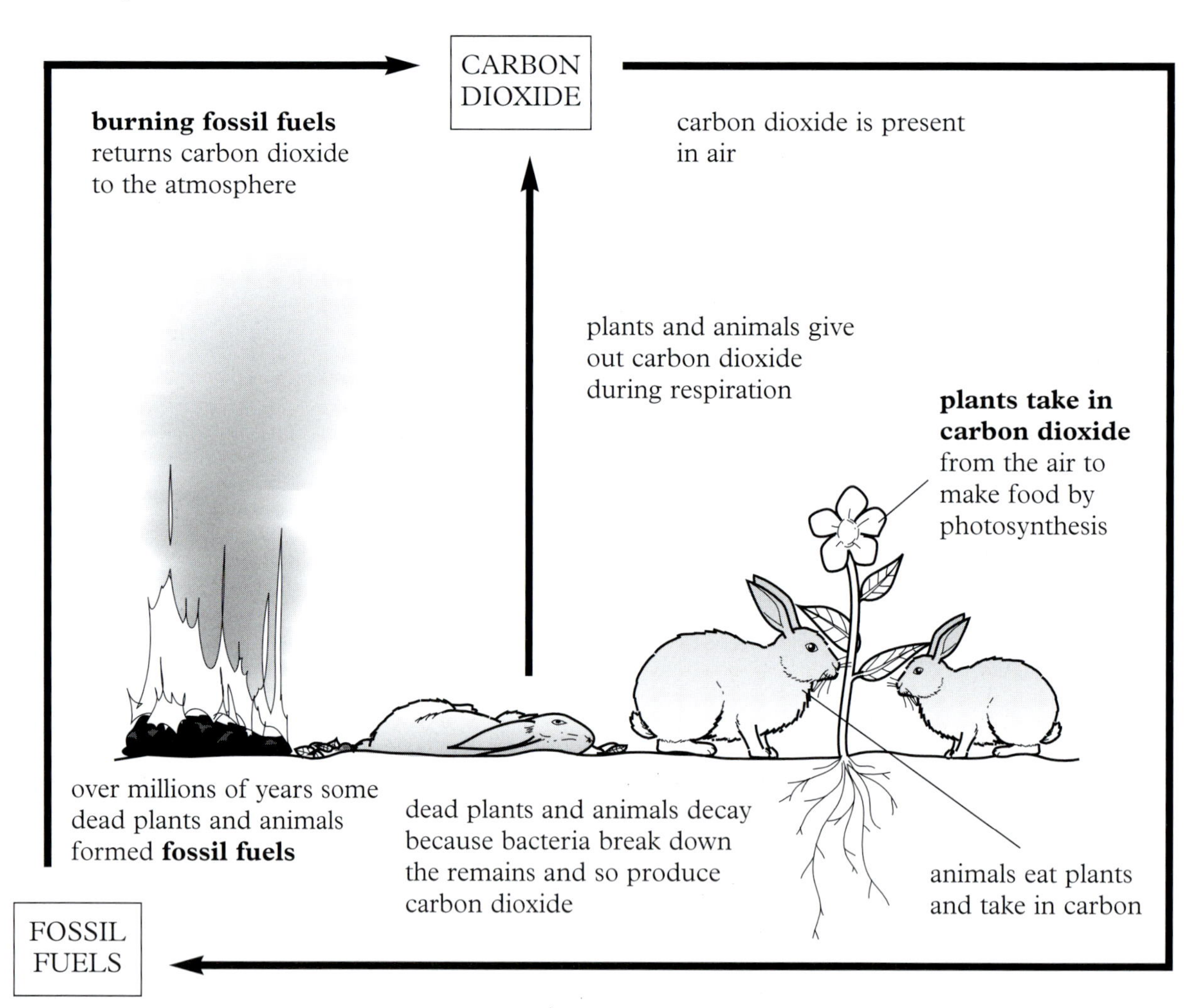

CARBON DIOXIDE
burning fossil fuels returns carbon dioxide to the atmosphere
carbon dioxide is present in air
plants and animals give out carbon dioxide during respiration
plants take in carbon dioxide from the air to make food by photosynthesis
over millions of years some dead plants and animals formed **fossil fuels**
dead plants and animals decay because bacteria break down the remains and so produce carbon dioxide
animals eat plants and take in carbon
FOSSIL FUELS

Questions

1. How is carbon taken out of the atmosphere?
2. How is carbon put back into the atmosphere?
3. Explain why the amount of carbon dioxide in the atmosphere stays about the same?
4. How is the plant in the diagram above 'locking up' carbon?

The greenhouse effect

Carbon dioxide is a useful gas to have in the air because it traps heat that would otherwise escape into space. Carbon dioxide acts like the glass in a greenhouse, which is why it is called a greenhouse gas. The Earth is warm enough to live on because of the greenhouse effect.

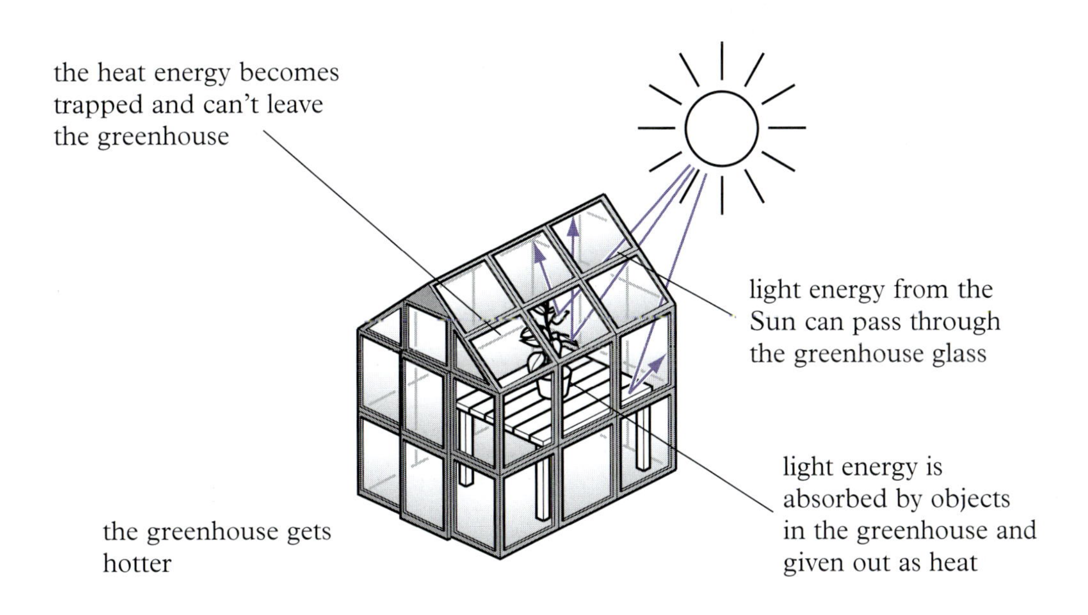
the heat energy becomes trapped and can't leave the greenhouse
light energy from the Sun can pass through the greenhouse glass
light energy is absorbed by objects in the greenhouse and given out as heat
the greenhouse gets hotter

As the level of carbon dioxide in the air increases, the Earth gets a little bit warmer. This process is called **global warming**.

The problem with burning

Most of the fuels we use, like natural gas, petrol and coal, are fossil fuels, formed from animals and plants that were once living. Living material contains carbon. When it burns, the carbon reacts with oxygen in the air to form carbon dioxide. The more fossil fuels we use the more carbon dioxide is put into the air.

Why global warming is a problem

The climate is very sensitive to small changes, so even a small rise in temperature could have a large effect on the planet.

If there is a rise in temperature flooding might happen because:

1 water expands when it is heated, which will make the oceans rise;
2 the polar ice caps would begin to melt, causing more water to be released into the oceans.

A rise in temperature could affect the weather in the following ways: deserts could become drier and wet areas wetter (crops might not grow under changed weather conditions). Storms and hurricanes could happen more often and with more force.

Questions

1 Name three fossil fuels.

2 Why might burning fossil fuels lead to an increase in the Earth's temperature.

Summary

The Earth's atmosphere millions of years ago was mainly made of carbon dioxide. Once the planet cooled, oceans formed and plants appeared in the water. Photosynthesis by the plants used up some carbon dioxide and produced oxygen.

A series of cycles keeps the mixture of gases in air roughly the same. The main cycles involve water, carbon and nitrogen. Burning fossil fuels produces carbon dioxide. Carbon dioxide is a greenhouse gas. The greenhouse effect could cause global warming.

Key words

fossil fuels Fuels formed by the decay of once-living material that has been laid down over millions of years.

global warming The way in which the Earth could warm, possibly causing changes in the weather.

greenhouse gas A gas that traps heat in the atmosphere.

photosynthesis The way in which plants use carbon dioxide and water to build carbohydrates, giving out oxygen as a waste product.

End of Chapter 8 questions

1 a) What is the ozone layer?
b) Why was ozone important for the beginning of life?
c) Why is ozone still important?

2 a) 'Carbon dioxide is a greenhouse gas'. Explain what this means.
b) Why is it very important to leave a window open if you leave a dog in a car on a warm day?
c) What might happen if the level of greenhouse gases in the atmosphere rises?
d) Suggest some ways in which we could stop this happening.
e) Make a poster about the possible dangers if we do not stop this happening.

3 a) What is meant by the 'water cycle'?
b) In the water cycle, water changes into water vapour. What is the name of this process?

c) Where does the energy come from for water to change into water vapour?
d) What happens to the water vapour as the air gets colder?
e) What is precipitation?

4 a) Why are bacteria important in the nitrogen cycle?
b) Ammonia is made by the Haber Process and is converted to fertilisers. What element is present in ammonia that is used by plants to make proteins?
c) How is nitrogen released back into the atmosphere?.

9 Rocks and the rock cycle

Most of the time we do not think about what lies underneath our feet, below the surface of the Earth. When volcanoes erupt or earthquakes shake the surface we are reminded about the Earth's structure. Below the outer layer of solid rock is thick partly-melted rock. This rock is like thick treacle. From time to time this melted rock bursts through the outer layer where we live. In fact, the inner Earth affects the outer layer all the time, but this takes place so slowly that we don't realise it is happening.

Earth's structure

The planet Earth is made of layers. It gets hotter and hotter the further we go into the centre. We live on the outer layer of the Earth's structure.

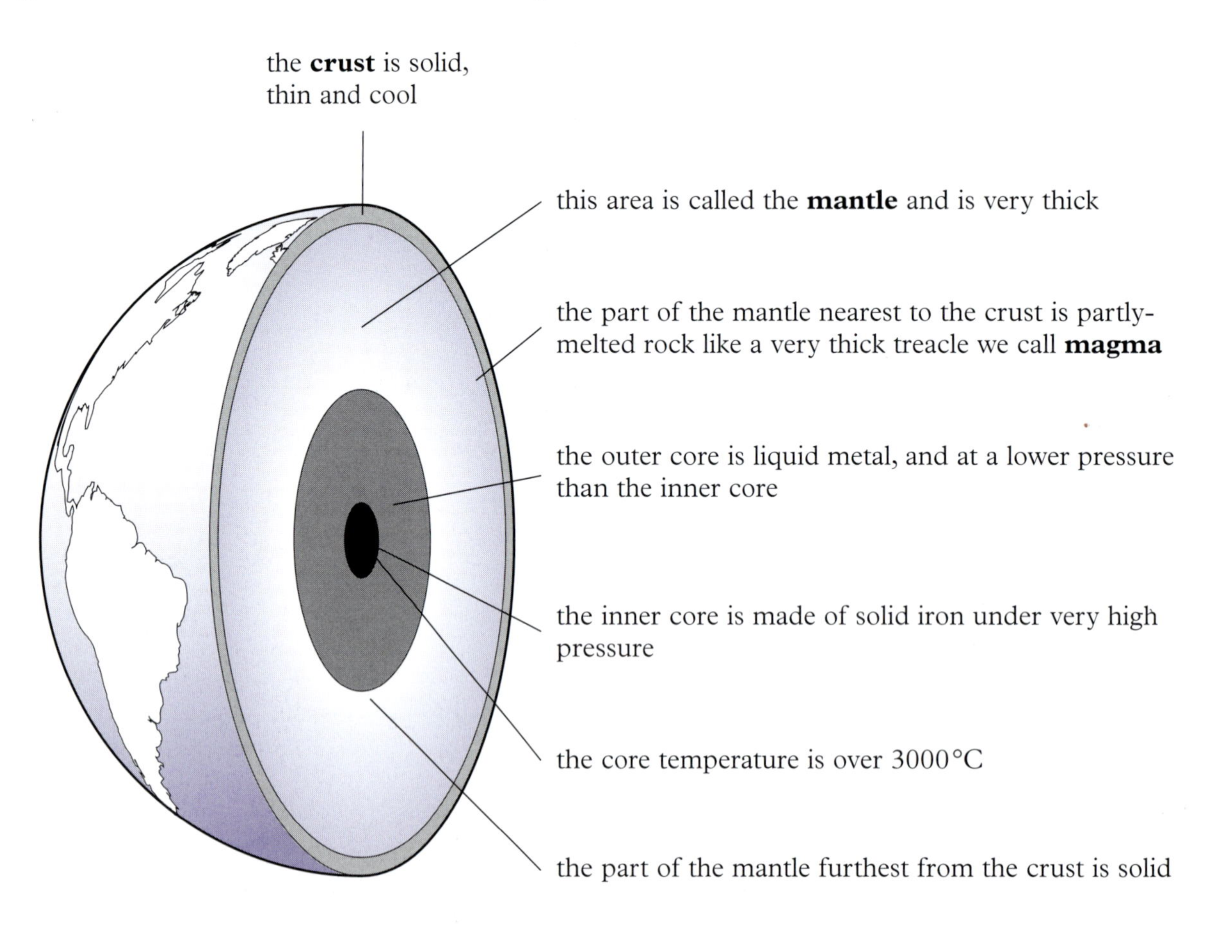

the **crust** is solid, thin and cool
this area is called the **mantle** and is very thick
the part of the mantle nearest to the crust is partly-melted rock like a very thick treacle we call **magma**
the outer core is liquid metal, and at a lower pressure than the inner core
the inner core is made of solid iron under very high pressure
the core temperature is over 3000 °C
the part of the mantle furthest from the crust is solid

The rock cycle

The outer crust has different land forms, oceans and mountains. Changes in the outer crust take place very slowly. Rocks are being made and worn away over millions of years. Changes are taking place all the time. We call this the **rock cycle**.

The rock cycle gives us three main types of rock:

igneous
sedimentary
metamorphic.

Millions of years ago only igneous rocks were present. Over time, some igneous rocks have been changed into sedimentary or metamorphic rocks.

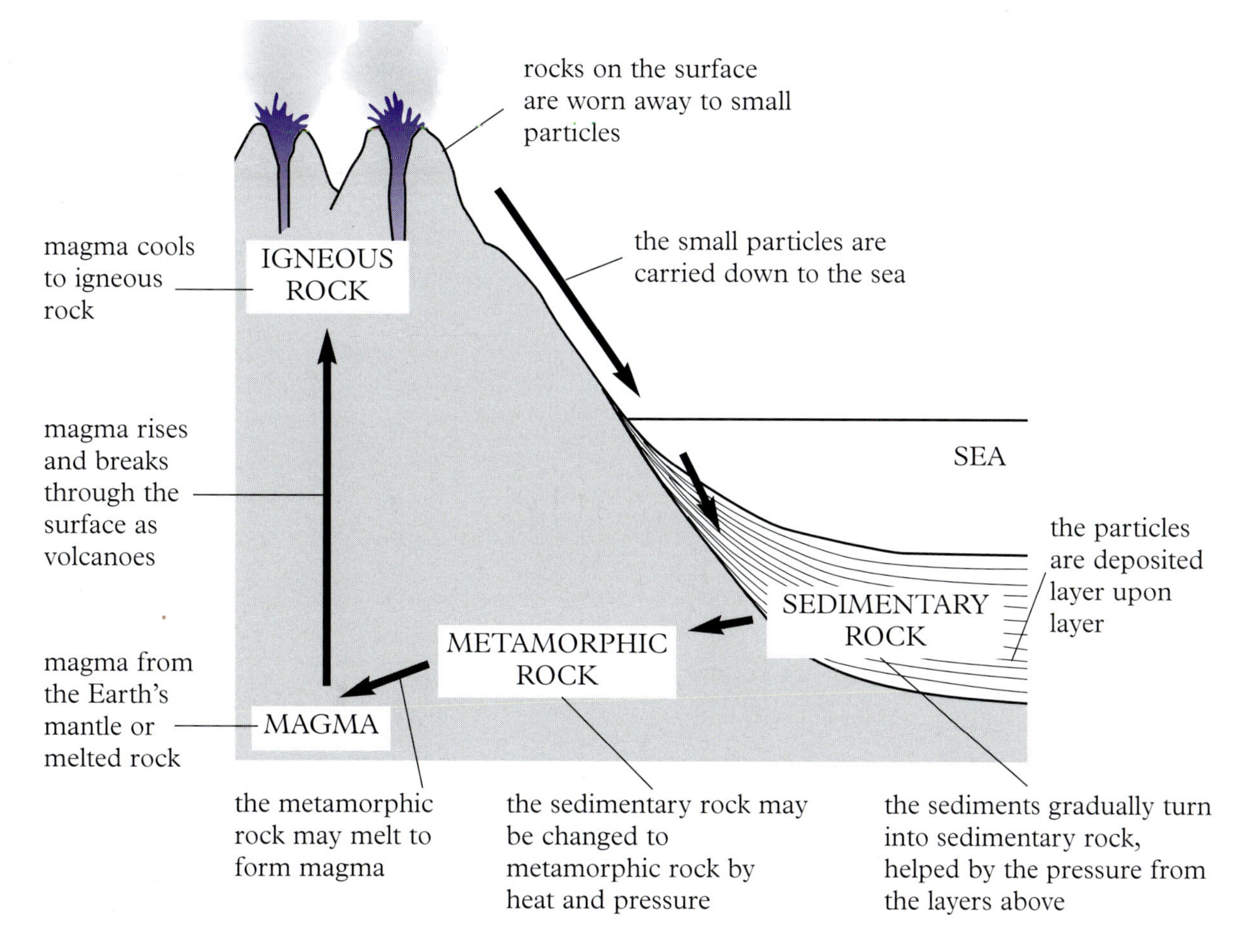
rocks on the surface are worn away to small particles
the small particles are carried down to the sea
magma cools to igneous rock
IGNEOUS ROCK
magma rises and breaks through the surface as volcanoes
SEA
the particles are deposited layer upon layer
SEDIMENTARY ROCK
METAMORPHIC ROCK
magma from the Earth's mantle or melted rock
MAGMA
the metamorphic rock may melt to form magma
the sedimentary rock may be changed to metamorphic rock by heat and pressure
the sediments gradually turn into sedimentary rock, helped by the pressure from the layers above

Some sedimentary rock may return to the surface due to movements of the Earth.

Questions

1 What is the name of the Earth's layer on which we live?

2 a) Where is magma found?
b) Describe what magma is.

3 Explain how the outer core and inner core differ.

4 Name the three main types of rock.

How rocks get worn away

Rocks are worn away by a process called **weathering**. There are three different ways of weathering rocks:

1 physical weathering
2 chemical weathering
3 biological weathering

Physical weathering

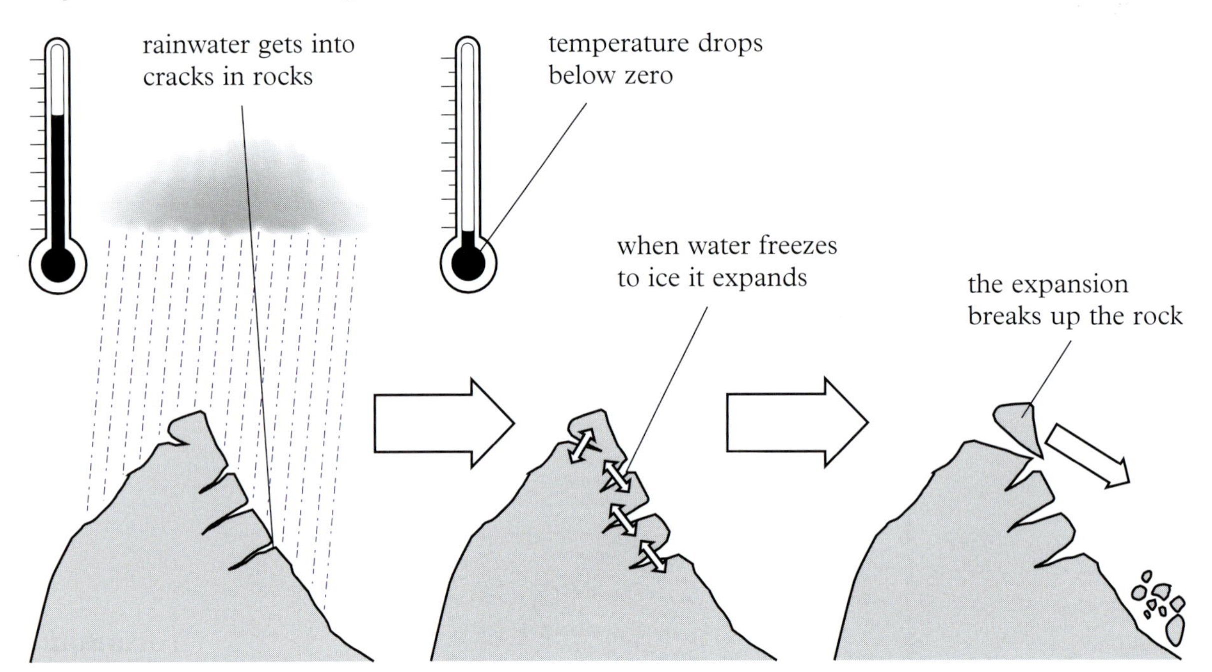

rainwater gets into cracks in rocks
temperature drops below zero
when water freezes to ice it expands
the expansion breaks up the rock

Chemical weathering

Rainwater is slightly acidic and can react with some types of rock.

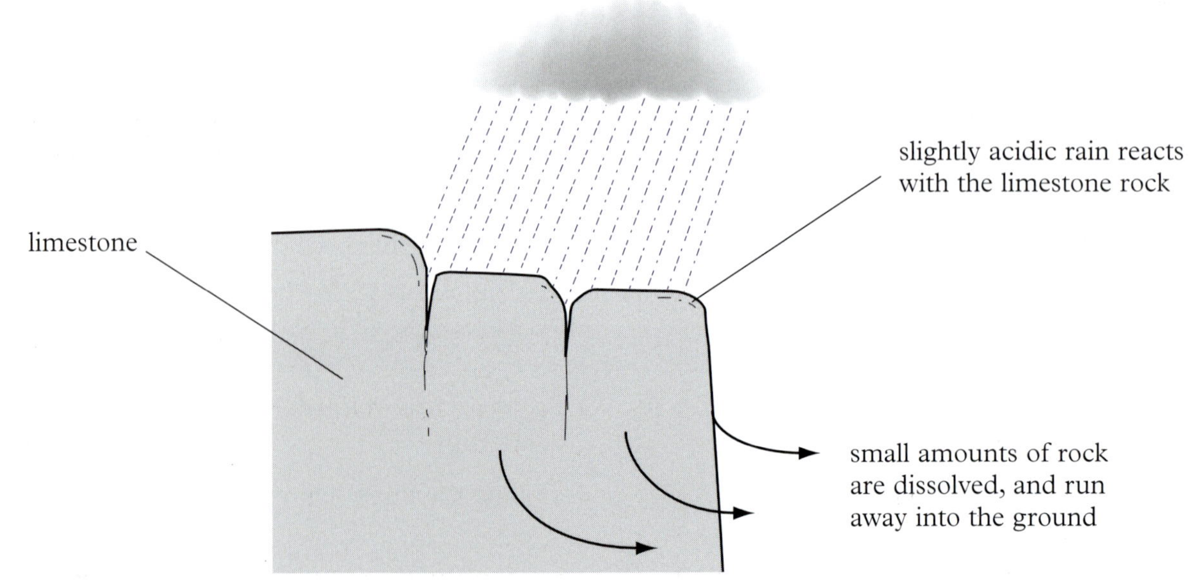

limestone
slightly acidic rain reacts with the limestone rock
small amounts of rock are dissolved, and run away into the ground

You can also see the effect of acid rain on buildings and statues that are made of limestone.

Biological weathering

Plants play a part in weathering rocks.

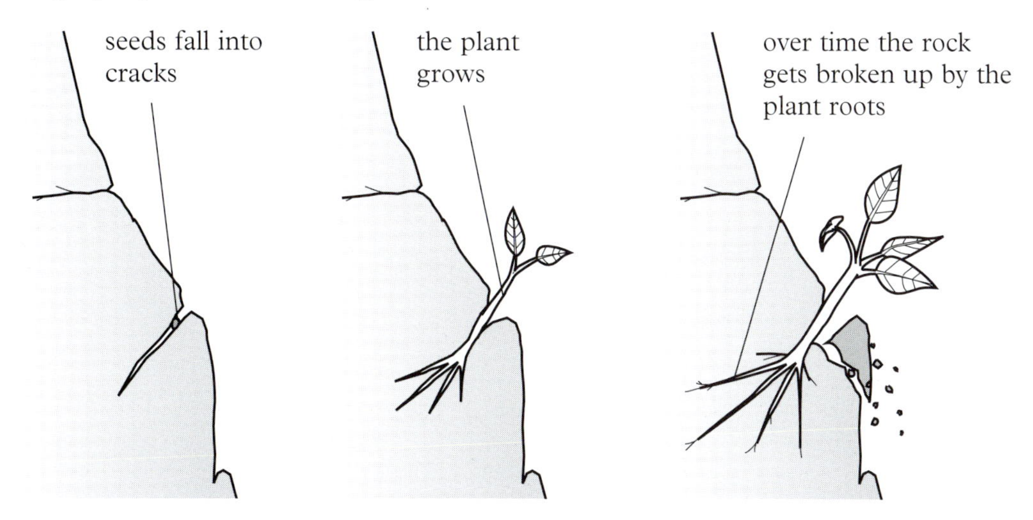
seeds fall into cracks
the plant grows
over time the rock gets broken up by the plant roots

Questions

1 a) Name three ways of weathering rocks.
  b) Which weathering process would take the longest to change rock into smaller particles.

2 On paving stones you can see the results of weathering by plants and the freeze-thaw of water. Explain how a paving stone might gradually be broken up by these sorts of weathering.

3 A plastic bottle was filled to the top with water and put in the freezer until the water froze. Draw a diagram to show what would happen if:
  a) there was no lid to the bottle – draw the result
  b) the bottle was tightly screwed shut?

Erosion

Rocks simply get worn away being constantly battered by smaller particles of rock. Wearing away rocks in this way is called **erosion**. Cliffs around the coast are constantly being eroded by the pounding of waves.

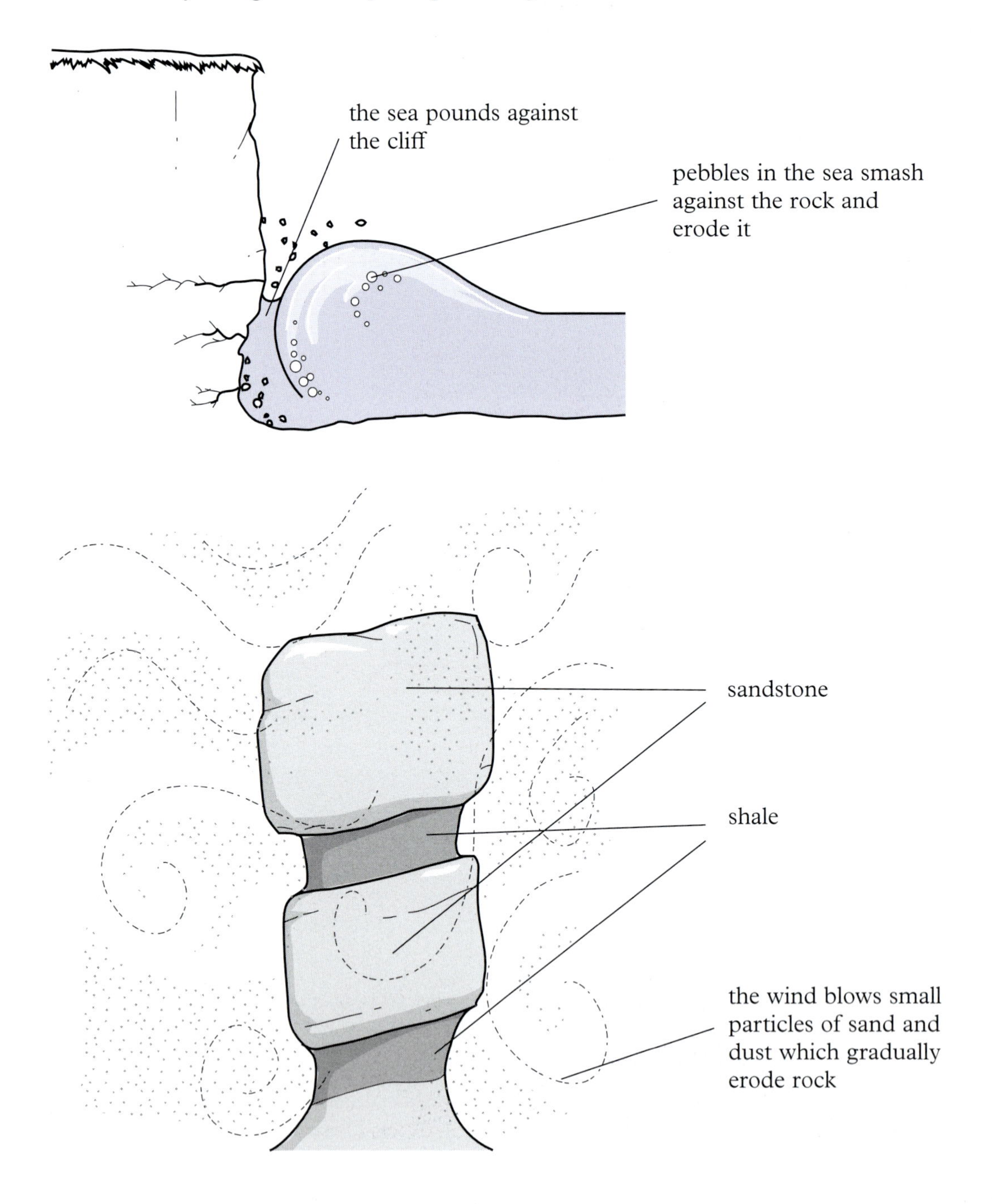

the sea pounds against the cliff
pebbles in the sea smash against the rock and erode it
sandstone
shale
the wind blows small particles of sand and dust which gradually erode rock

Questions

1 The boulder in the diagram above has eroded away.

a) Which type of rock is softer? Explain your answer.

b) What caused the erosion?

The particles of rocks are moved

The movement of the rock particles that have been worn away is called **transport**.

Gravity means that any particles that have been eroded always fall downwards. If the particles end up in flowing water then they are carried or transported along with it.

The different types of rock

Igneous rock

Igneous rocks are formed during volcanic action. Melted rock known as magma bursts up through the mantle when a volcano erupts.

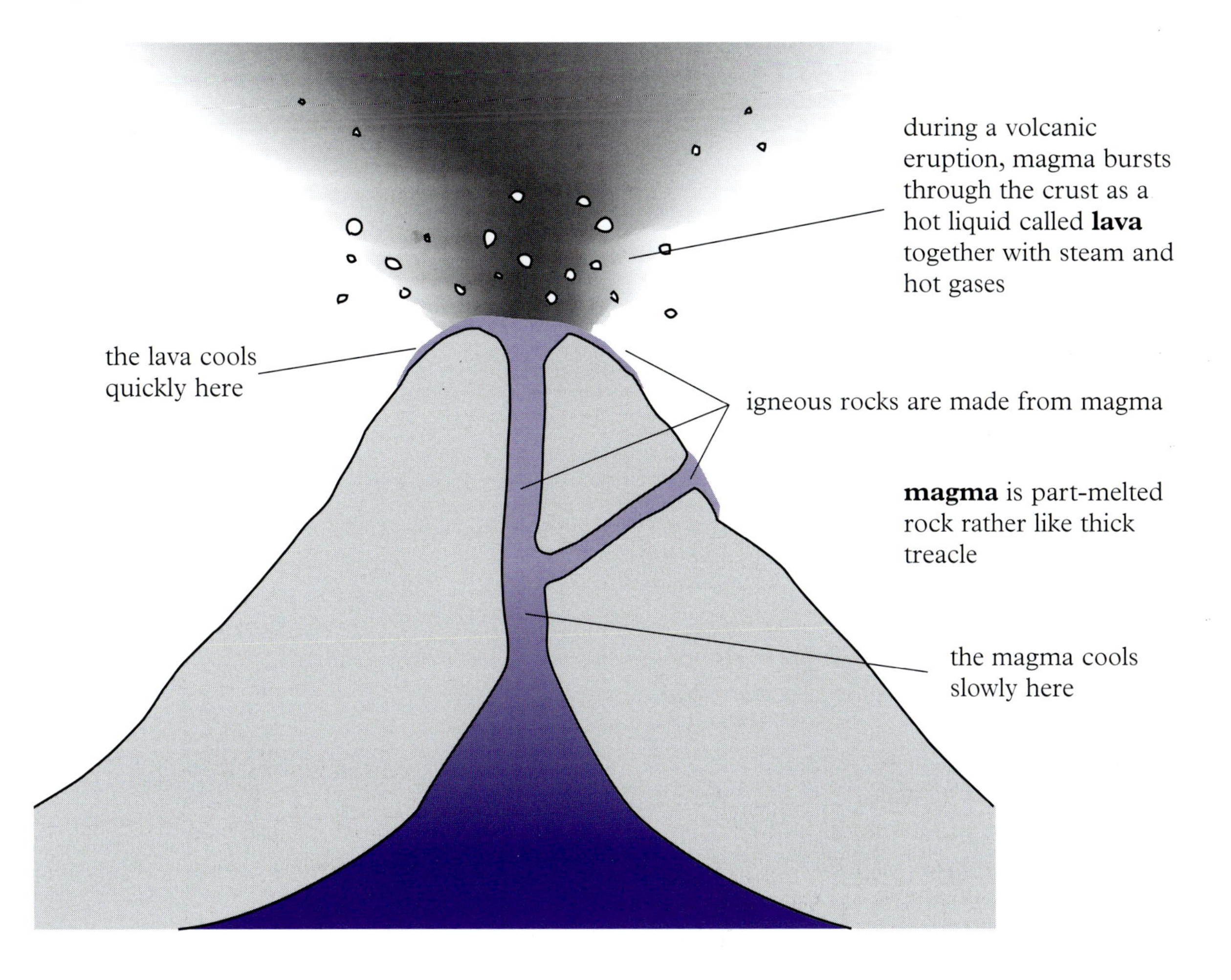
during a volcanic eruption, magma bursts through the crust as a hot liquid called **lava** together with steam and hot gases
the lava cools quickly here
igneous rocks are made from magma
magma is part-melted rock rather like thick treacle
the magma cools slowly here

When the volcano stops erupting, the magma cools and becomes igneous rock. The type of igneous rock formed depends on the speed at which the magma cools.

If the liquid magma **cools slowly**, the rock will be made up of **large crystals**.

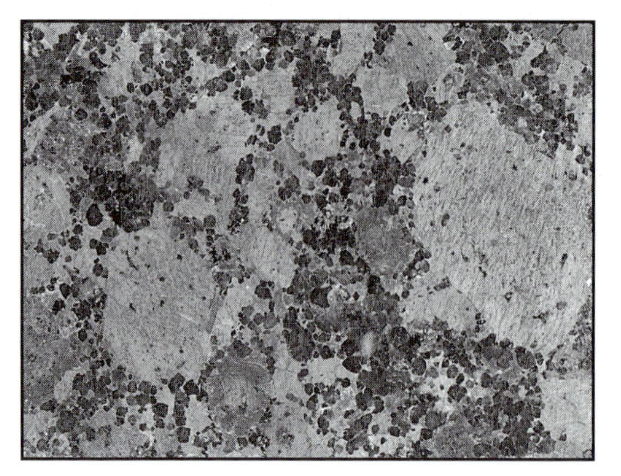

Granite is a typical igneous rock with large crystals. Granite is formed when the liquid magma cools slowly **deep inside** the Earth. Magma cools more slowly inside the Earth because it is much warmer there.

If the liquid magma **cools quickly**, the rock will be made up of **small crystals** or even have a glassy look.

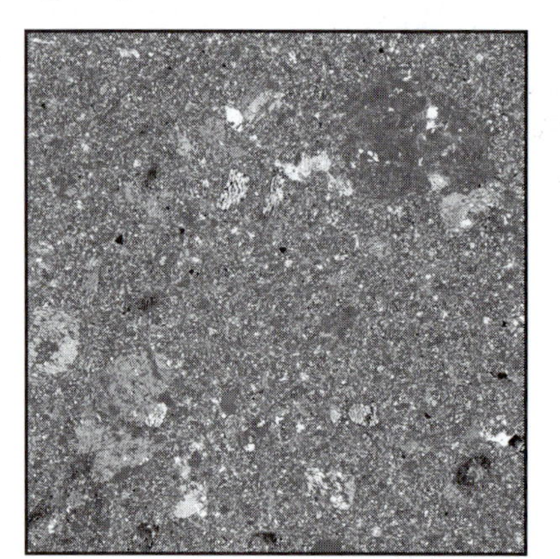

Basalt is an igneous rock with very small, dark crystals. It is formed when the rock cools **at the surface** of the Earth. Magma cools much more quickly when it is in contact with air at the surface of the Earth.

Questions

1 Why would you expect liquid rock to cool more quickly outside the Earth compared to deep inside the Earth? Use the idea of a hot drink left outside to help your explanation.

2 a) In the diagram put the rocks at A, B and C in order of cooling, beginning with the one that cooled the slowest.

b) What is the difference in the size of the crystals of each rock?

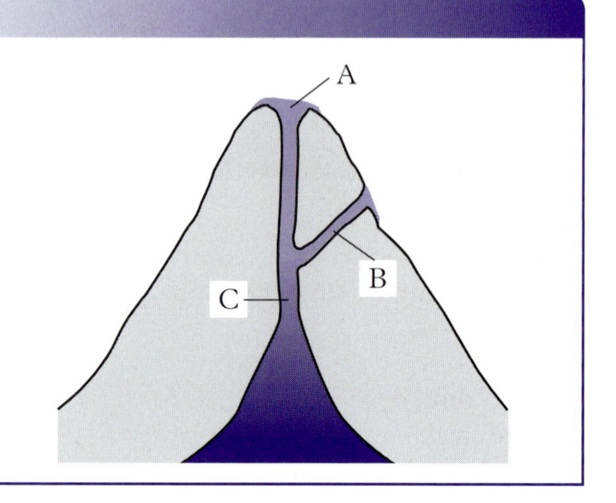
A
B
C

Sedimentary rocks

Most sedimentary rocks are made from particles which fall layer upon layer over millions of years. All of these particles came from older rocks that have been slowly worn away.

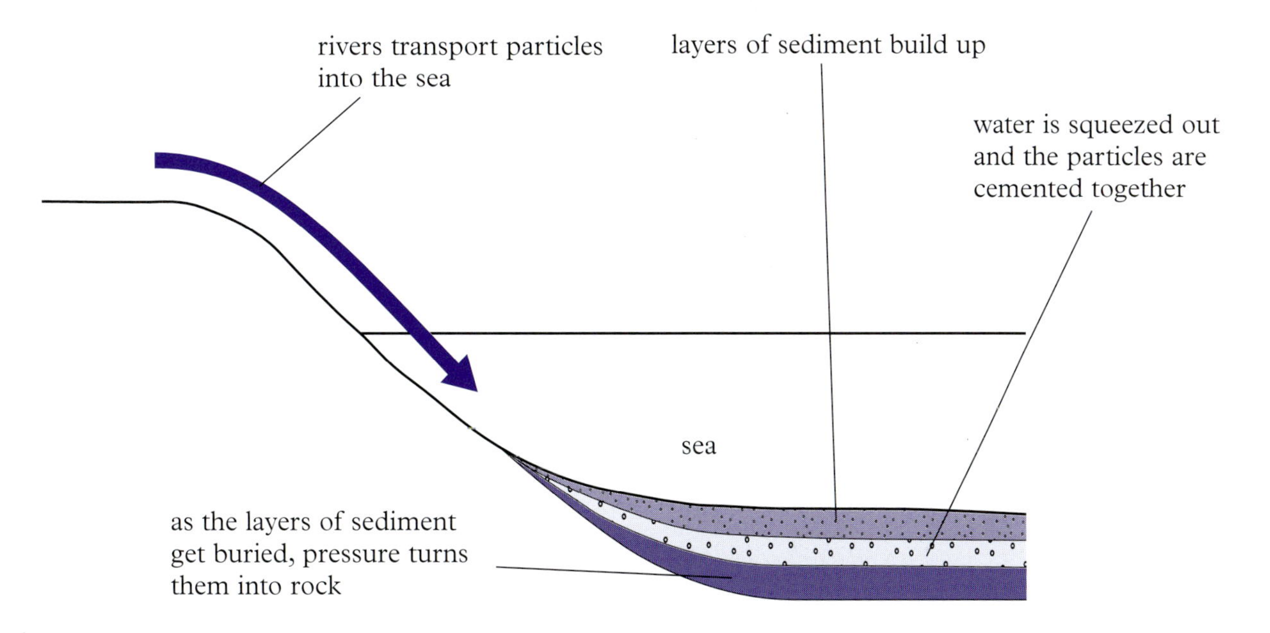
rivers transport particles into the sea
layers of sediment build up
water is squeezed out and the particles are cemented together
sea
as the layers of sediment get buried, pressure turns them into rock

There are different sorts of sedimentary rock. The type of rock depends on the sediment that it is made from.

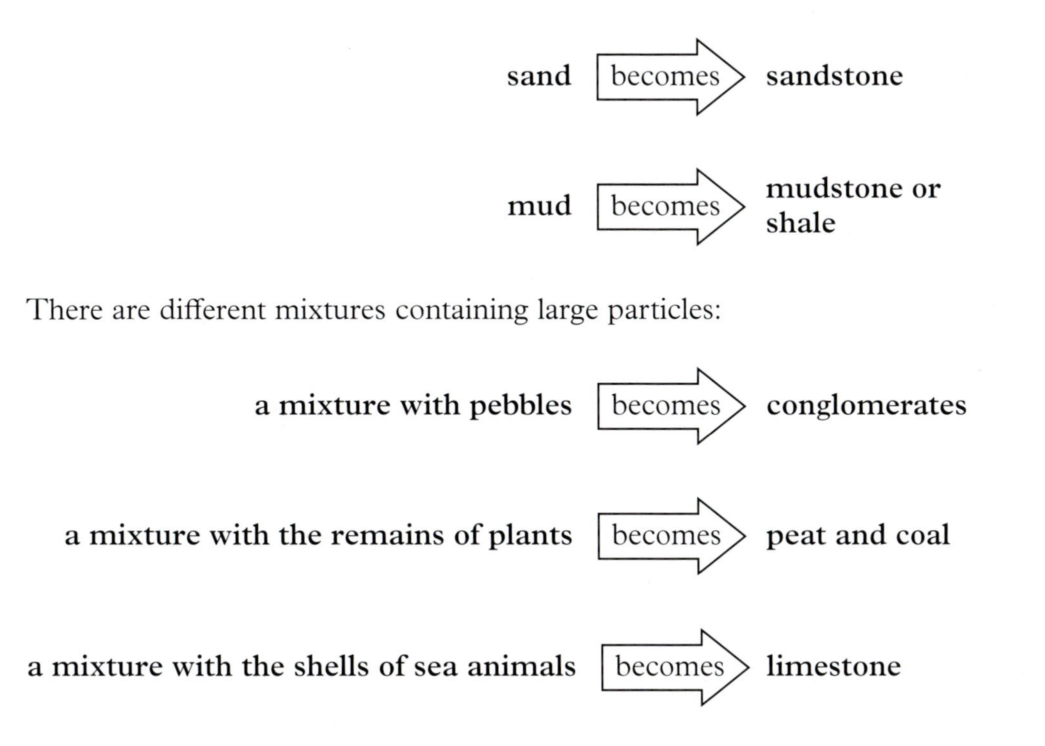
sand
becomes
sandstone
mud
becomes
mudstone or shale
There are different mixtures containing large particles:
a mixture with pebbles
becomes
conglomerates
a mixture with the remains of plants
becomes
peat and coal
a mixture with the shells of sea animals
becomes
limestone

Some sedimentary rocks form from the evaporation of water. Rock salt was formed in this way. When the sea water evaporated away over millions of years rock salt was left behind.

Fossils in sedimentary rocks

As layer upon layer of rock particles was laid down, some dead animals or plants were completely buried in deep layers. If this happened quickly, before the plant or animal could rot away, they may have been preserved in some way. When this happened **fossils** were formed. Fossils were formed when a plant or animal was preserved in one of the following ways:

1. the original plant may still be there
2. the material which made up the plant or animal could have been replaced by a mineral
3. the plant or animal may have rotted away and left a hollow imprint.

this is an example of a fossil imprint

the animal that left this imprint died millions of years ago

the body of the animal rotted away

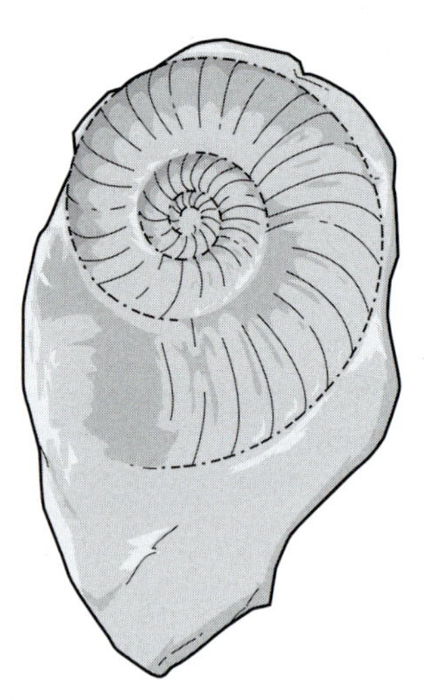

Fossils have been very useful in finding out how old a rock is because some animals and plants only lived at a particular time in history. The time line below shows the dates that some creatures first appeared.

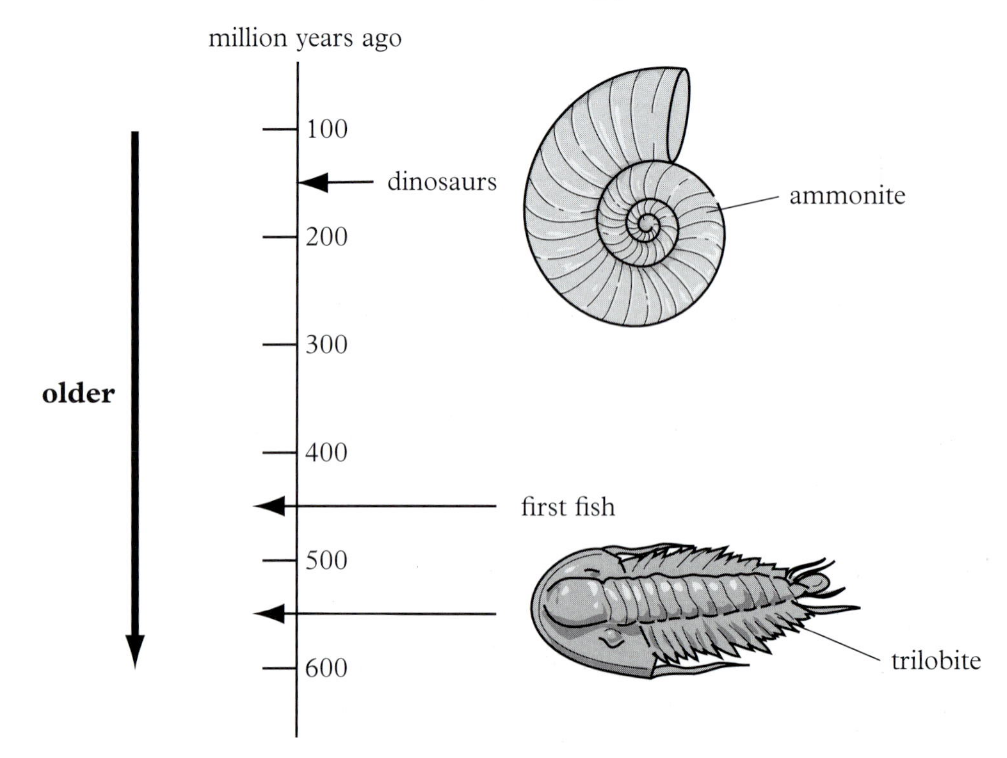
million years ago
100
dinosaurs
200
300
older
400
first fish
500
600
ammonite
trilobite

Knowing when fossils were formed also means that we can match up rocks of the same age that are not now at the same depth.

the fossils in layer A match with the fossils in layer B which means the rocks in layers A and B are of the same age

A

B

How to recognise sedimentary rocks

- They may have **powdery surfaces**.
- You can often see the **layers** that make them up.
- They might contain **fossils**.
- Chalk and limestone are made up of the chemical calcium carbonate. Calcium carbonate gives off the gas carbon dioxide when it reacts with an acid. A test for chalk or limestone is to add an acid to a rock sample and see if it fizzes. You could try this yourself using vinegar as the acid.

Metamorphic rocks

Metamorphic rocks are rocks that have been changed by great heat and pressure under the ground. They may originally have been sedimentary rocks, igneous rocks or even metamorphic rocks themselves. The rocks are not melted but their structure changes. For example:

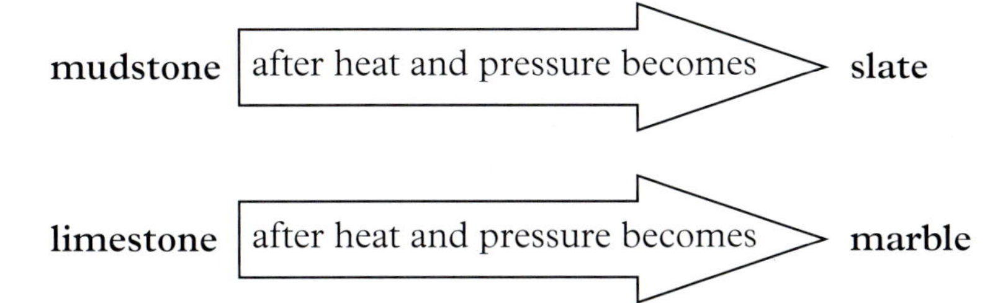
mudstone
after heat and pressure becomes
slate
limestone
after heat and pressure becomes
marble

Rocks often become harder and more crystalline after metamorphosis.

For example, mudstone turns to a rock called gneiss under more extreme heat and pressure. This is the pink 'streaky bacon' rock.

Metamorphosis due to deep burial

Rocks can be changed into harder rock by pressure and very high temperatures. When rock changes in this way it is called **metamorphosis**.

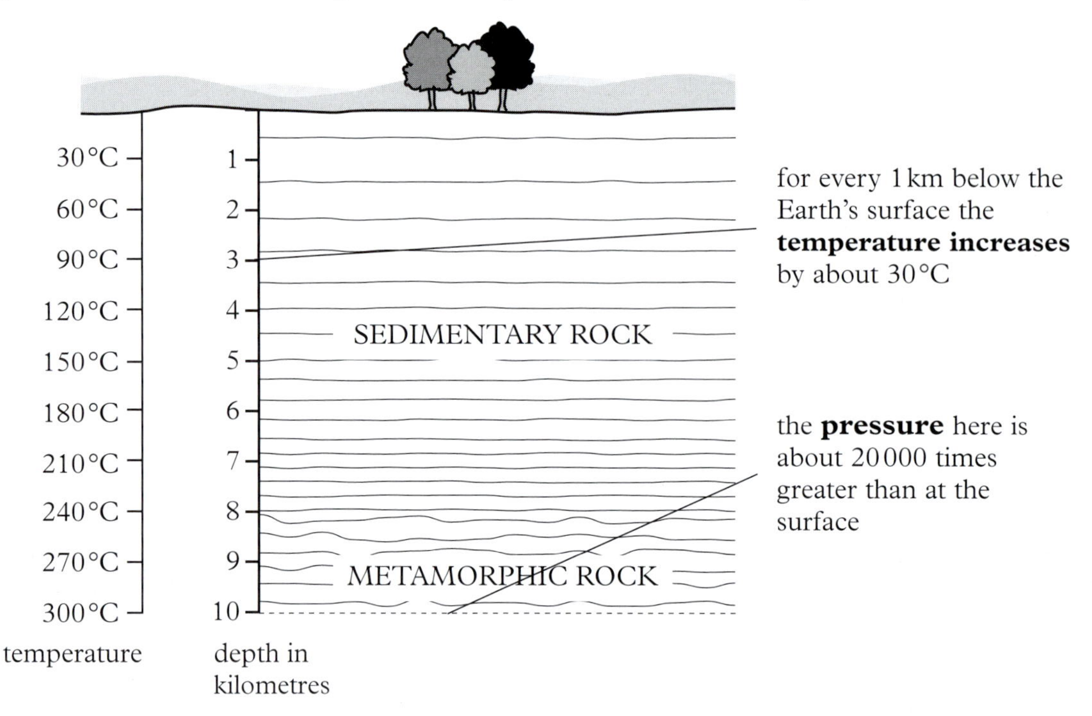
30 °C
60 °C
90 °C
120 °C
150 °C
180 °C
210 °C
240 °C
270 °C
300 °C
temperature
1
2
3
4
5
6
7
8
9
10
depth in kilometres
SEDIMENTARY ROCK
METAMORPHIC ROCK
for every 1 km below the Earth's surface the **temperature increases** by about 30 °C
the **pressure** here is about 20 000 times greater than at the surface

Metamorphosis due to the heat from a volcano

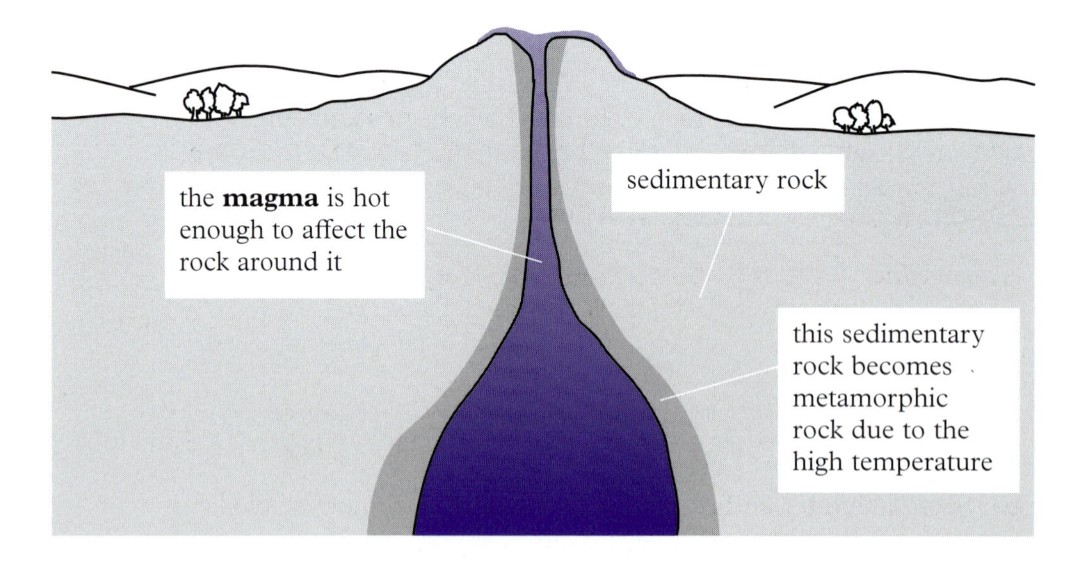
the **magma** is hot enough to affect the rock around it
sedimentary rock
this sedimentary rock becomes metamorphic rock due to the high temperature

Questions

1 What two things can change a rock into a metamorphic rock?

2 Explain why metamorphosis of rock is likely to happen deep below ground rather than on the surface.

Summary

Earth has the core at its centre, surrounded by the mantle. The crust is on the outside. The Earth gets hotter and hotter as we go towards the centre.

Rocks are made and destroyed in the rock cycle. The rock cycle takes place over millions of years. Weathering and erosion cause rocks to be worn away.

There are three types of rock:

1 Sedimentary rock is made from particles of other rocks. This type of rock has many layers. Fossils may be found in sedimentary rocks.

2 Igneous rock is made from cooled magma. If the magma cooled quickly the rock has small crystals. If the magma cooled slowly the rock has large crystals.

3 Metamorphic rock is any rock that has been changed by extreme heat and pressure.

Fossils can help to tell us the age of a rock.

Key words

erosion The term used when rocks are worn away either by waves, pieces of rock hitting other rocks in river beds or by wind carrying sand that scours away soft rock.

fossils The remains of plants and animals that were buried deeply long ago so that they became part of the rock.

magma Partly-melted rock.

weathering The term for the breakdown of rocks by acidic rain water dissolving rocks, freezing and thawing of rain water in cracks and plants growing in cracks.

End of Chapter 9 questions

1 Fill in the blanks in the passage below about the rock cycle by using words from the list in the most suitable places. Each word may be used once, more than once or not at all.

metamorphic	**sedimentary**	**igneous**	**magma**	**small**	**large**
volcano	**lava**	**pressure**	**water**	**erosion**	**mantle**

Below the Earth's crust is the m__________ which is a layer of semi-molten rock. During the eruption of a volcano, this rock breaks through to the surface as a liquid called l__________ . When it cools it forms solid i__________ rocks. Rocks with different sized crystals are produced with different rates of cooling. Slow cooling, under the surface, produces rock such as granite, which has l__________ crystals. Existing rock may change with heat and p__________ , to a new type or rock called m__________ . All types of rock may be worn away by the action of wind and w__________ .

This process is called e__________ . The particles which have worn away may be deposited and later form s__________ rock.

2 The diagram below shows one way in which rocks are weathered.

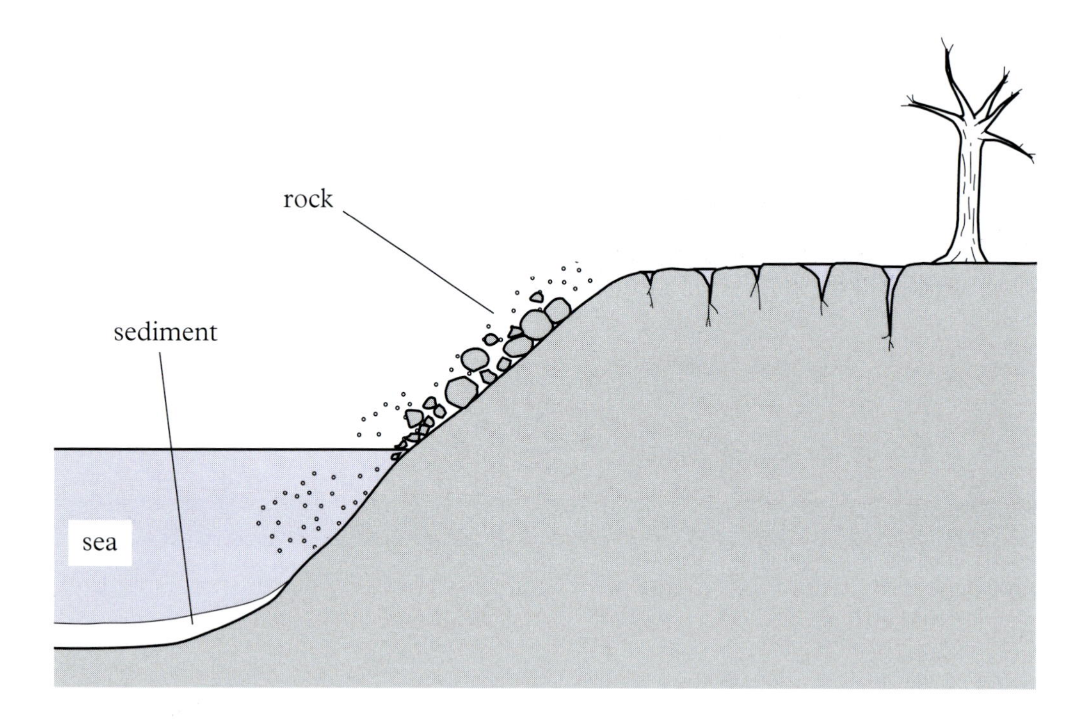
rock
sediment
sea

a) Explain how the rainwater in the cracks helps to break the rock up.
b) State two other ways in which rock is broken down into small particles.
c) Over time, the sediment may become changed into rock. What is the name of this type of rock?

3 a) One of the three main types of rock is called **igneous**.
b) What are the other two main types of rock called?
c) Some types of rock contain fossils. What is a fossil?
d) Which of the three main types of rock cannot contain fossils? Explain your answer.

4 a) Use the words below to answer this part of the question.

granite basalt shale slate sandstone limestone marble

i) Name a metamorphic rock.
ii) Name an igneous rock.
iii) Name a sedimentary rock.

b) Igneous rocks are formed when magma cools.
i) What is magma?
ii) Explain why some igneous rocks have very small crystals and some have larger crystals.

c) Where on the diagram would you expect to find metamorphic rock? Explain your answer.

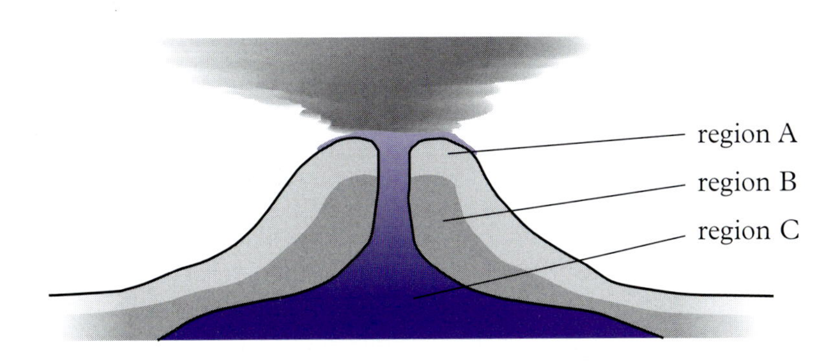
region A
region B
region C

10 The story of Earth

The theory of **plate tectonics** says that the Earth's surface is made up of rafts of solid rock, called plates. The plates float on a bed of treacle-like, half-melted rock. Using this idea and evidence from the rocks, Earth scientists have an idea of the history of the Earth's surface. This is the present day theory of how it all began.

About 250 million years ago there was just one giant piece of land above sea level. Scientists have given this the name **Pangaea**. Over millions of years Pangaea broke up and the pieces drifted apart.

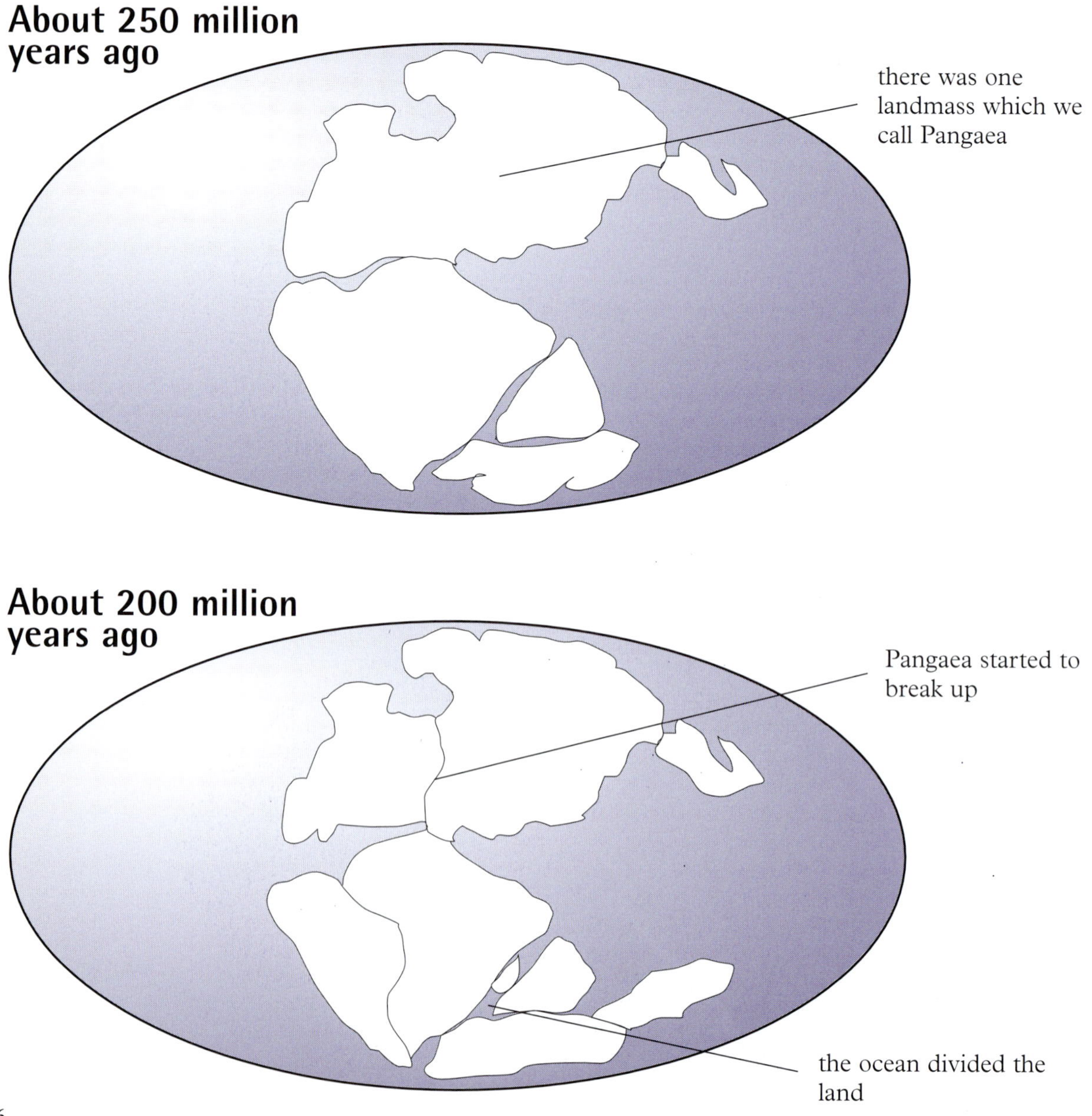
About 250 million years ago
there was one landmass which we call Pangaea
About 200 million years ago
Pangaea started to break up
the ocean divided the land

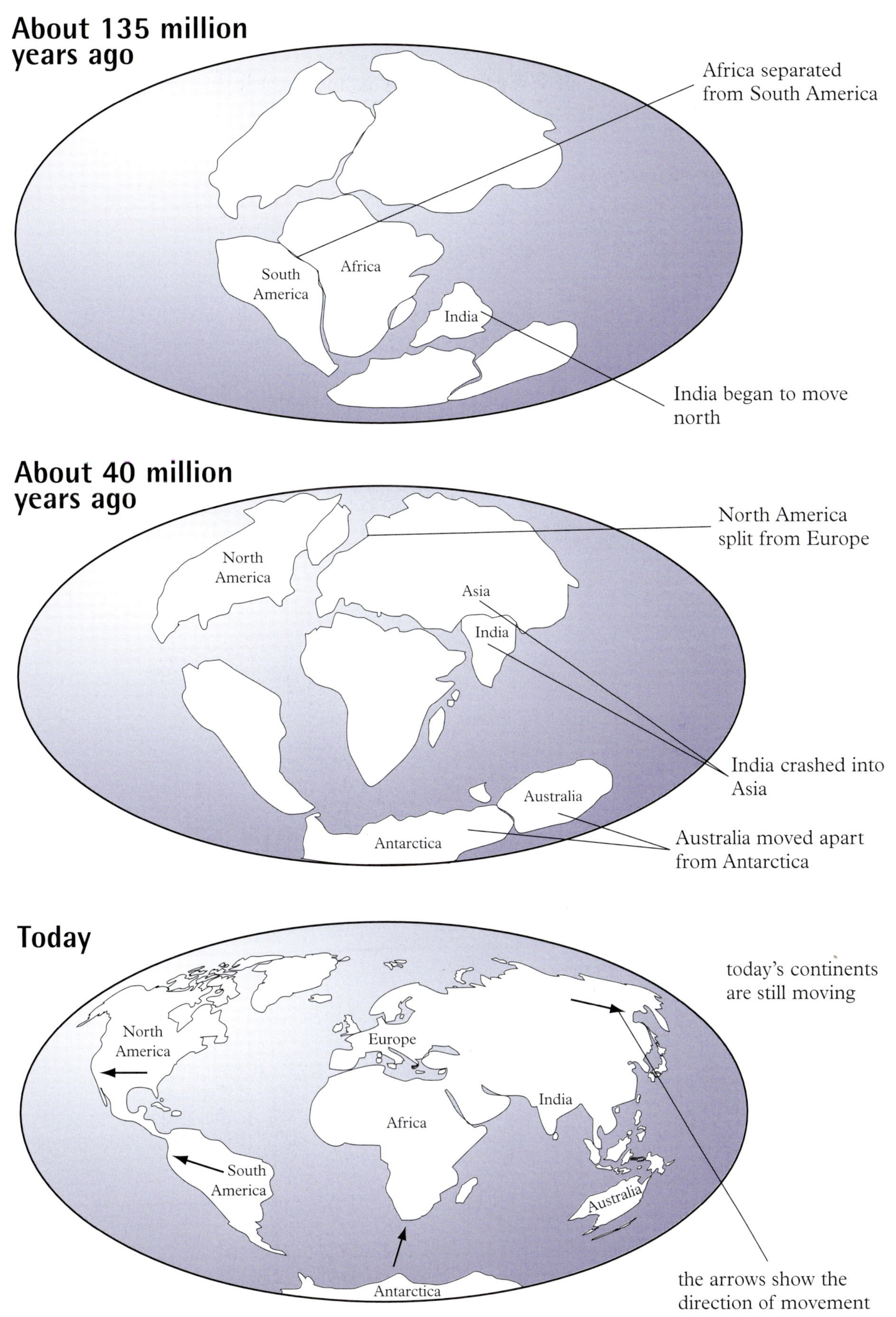
About 135 million years ago
Africa separated from South America
South America
Africa
India
India began to move north
About 40 million years ago
North America split from Europe
North America
Asia
India
India crashed into Asia
Australia
Antarctica
Australia moved apart from Antarctica
Today
today's continents are still moving
North America
Europe
India
Africa
South America
Australia
Antarctica
the arrows show the direction of movement

Evidence for Pangaea

The shapes of the continents we have today can be fitted together like pieces of a jigsaw. The shapes do not fit exactly but they do show how changes could have taken place.

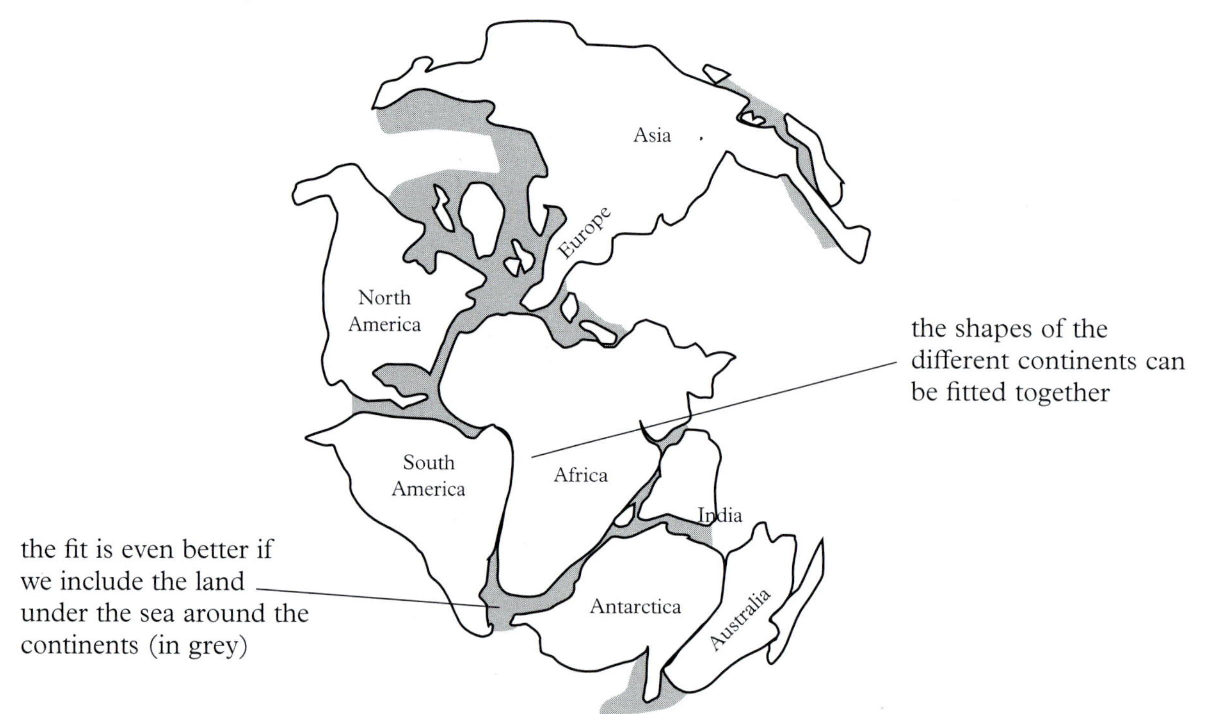
Asia
Europe
North America
South America
Africa
India
Antarctica
Australia
the shapes of the different continents can be fitted together
the fit is even better if we include the land under the sea around the continents (in grey)

How rock types support the idea of Pangaea

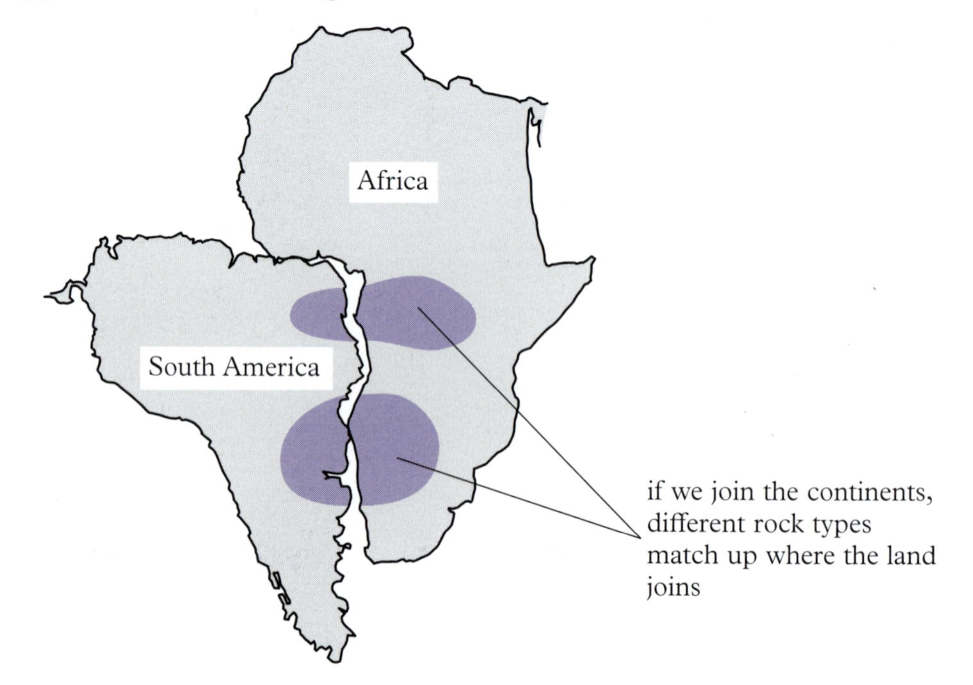
Africa
South America
if we join the continents, different rock types match up where the land joins

The ocean floor between the continents is thinner and newer than the main land masses. This means that there must have been a time when there was no ocean floor between continents.

The fossils found in the rocks on different continents are very similar – which suggests that the continents were once connected. For example, Lystosaurus was a reptile that lived in the hot climate of Africa, India and China 200 million years ago. Its fossil remains have also been found in Antartica.

Summary

The continents of the Earth were once joined together. Movement of the plates separated the continents, which caused the continents to drift.

Much evidence supports this. For instance, the shapes of the continents suggest they were once joined, while the rocks and the fossils on different continents match.

Key words

continents Large land masses.

fossils The remains of plants and animals that were buried deeply long ago so that they became part of the rocks.

End of Chapter 10 questions

1 Trace the shapes of the continents in the diagram below, cut them out, and then fit them together. Explain how we believe the continents came to be apart.

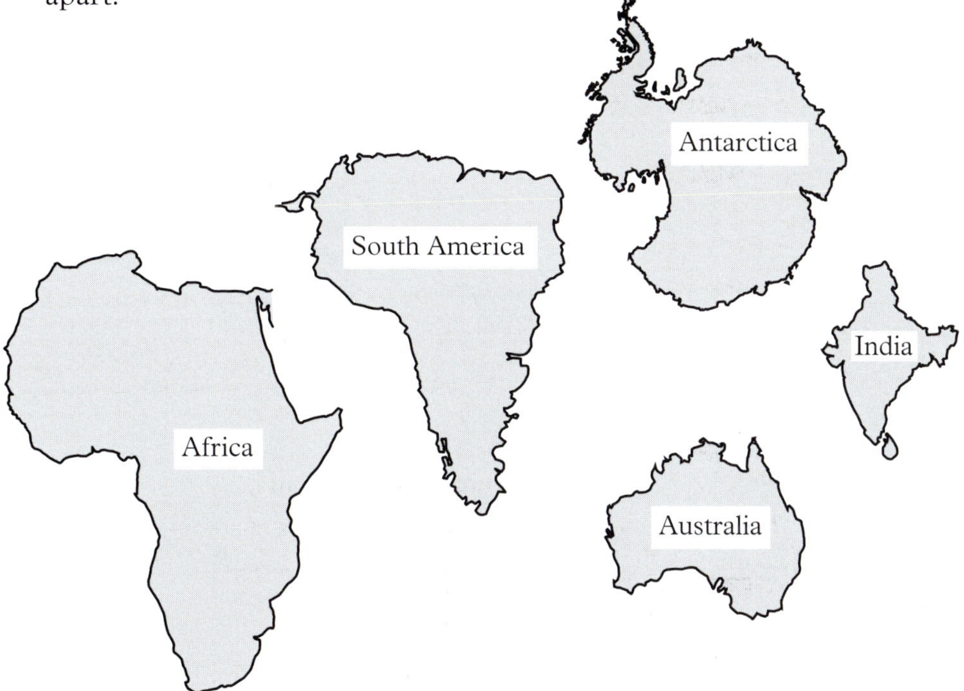
Antarctica
South America
Africa
India
Australia

2 The rocks in Northern Europe contain fossils that belong to a tropical climate. Explain this in terms of the idea that the continents are drifting.

11 About metals

Metals are very useful to us because of their special properties. Metals are **elements**, not compounds. This means that all the atoms in a metal are the same. Sometimes metals are mixed to form alloys like brass, steel or nickel-silver. If you look at some of the **physical properties** of metals you can see that they are alike in many ways:

Comparing the physical properties of metals and non-metals

The **physical properties** of a material are the ways that it behaves without the breaking of any chemical bonds to make a new substance.

Metal

hard and shiny solids

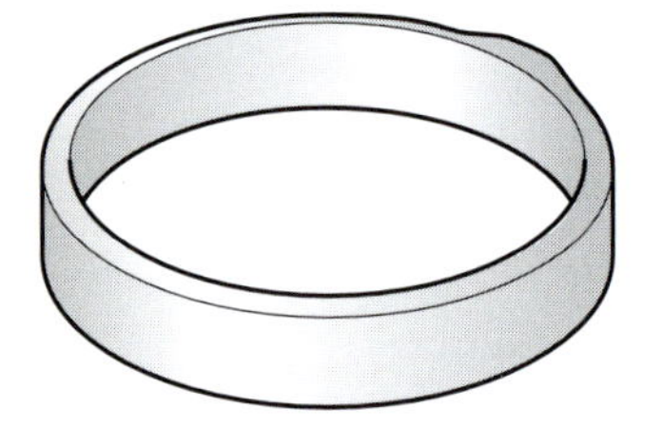

Non-metal

dull solids, or liquids or gases

good conductors of electricity

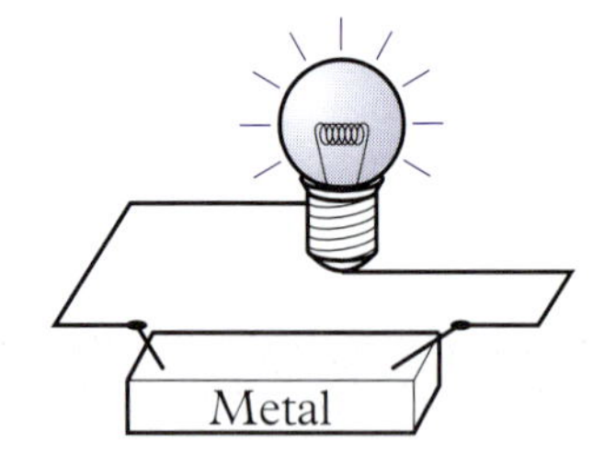
Metal

poor conductors of electricity

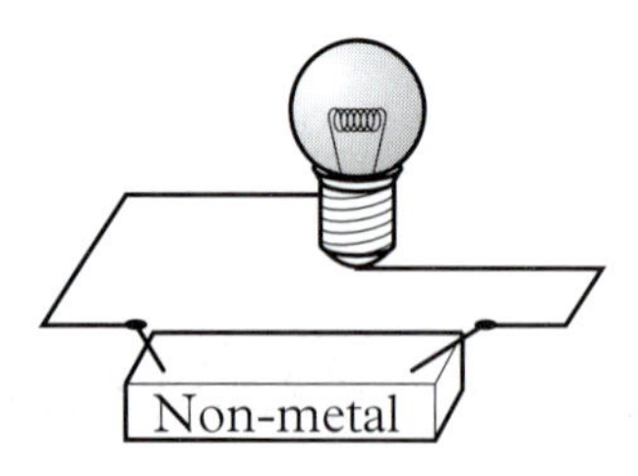
Non-metal

good conductors of heat

metal handle

poor conductors of heat

non-metal handle

Metal

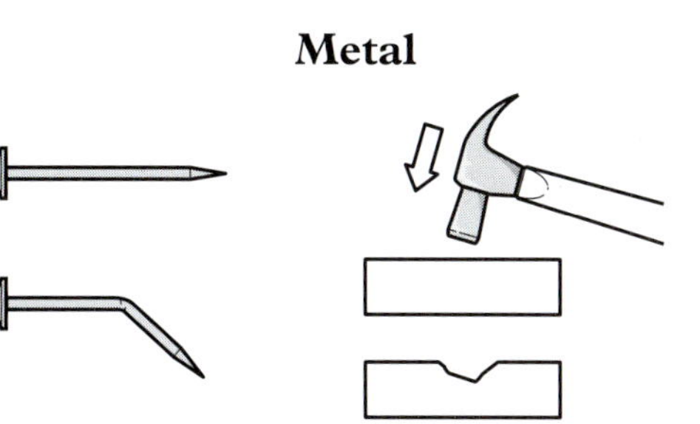

malleable which means that the metal **can be bent** and **hammered**

Non-metal

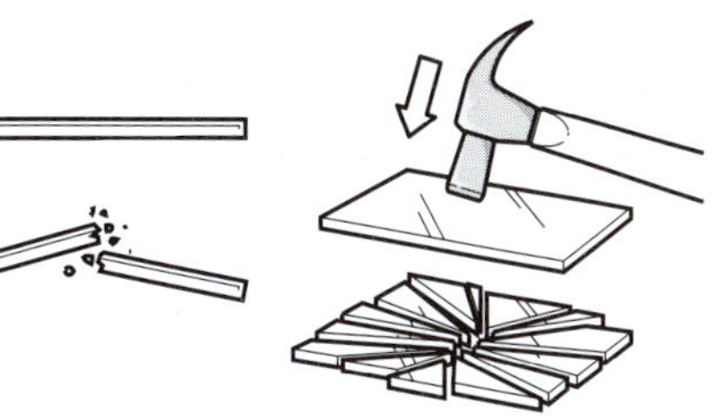

brittle which means that the non-metal **will snap when bent** and **shatter when hammered**

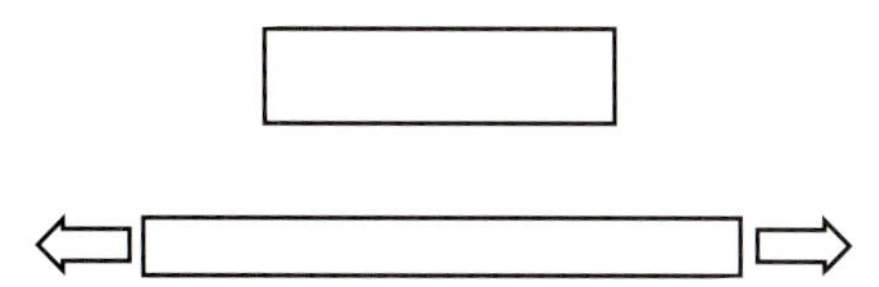

ductile which means that the metal **will stretch if pulled**

brittle which means that the non-metal **will snap if pulled**

Magnetism

Non-metals are not magnetic **and most metals are not magnetic** either. In fact **iron, cobalt and nickel** are the only **magnetic metals**.

The chemical reactions of metals

A typical metal is a hard and shiny solid which is a good conductor of heat and electricity. There are about eighty different metal elements. Some are very alike but no two are exactly the same. The **chemical** properties of a metal are the way it behaves when chemical bonds are broken and new substances are made.

The reaction of metals with oxygen in the air

Metals burn to form metal oxides.

Metal	Reaction with air	Does the oxide dissolve in water?
Copper	needs to be heated very strongly for some time before it forms a black coating of **copper oxide** on its surface	copper oxide **doesn't dissolve** in water
Magnesium	quickly burns away with a bright light to form **magnesium oxide**	**some** magnesium oxide **dissolves** in water to form an alkaline solution of **magnesium hydroxide**
Calcium	reacts even more vigorously than magnesium. It burns away completely with a brilliant crimson flame to form **calcium oxide**	calcium oxide **dissolves** in water to form a very alkaline solution of **calcium hydroxide**

The reaction of metals with cold water

Most metals do not react with cold water, or do so very slowly. Only very reactive metals react with cold water and release hydrogen from the water.

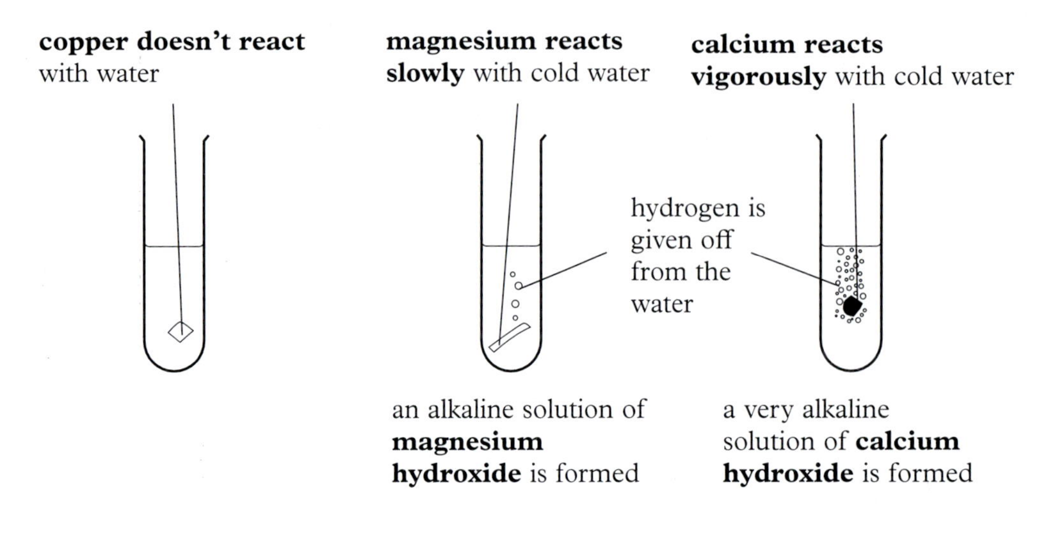

copper doesn't react with water
magnesium reacts slowly with cold water
calcium reacts vigorously with cold water
hydrogen is given off from the water
an alkaline solution of **magnesium hydroxide** is formed
a very alkaline solution of **calcium hydroxide** is formed

The reaction of metals with dilute acids

Reactive metals react with dilute acids to form a salt and hydrogen.

reactive metal + dilute acid ⟶ salt + hydrogen

magnesium + sulphuric acid ⟶ magnesium sulphate + hydrogen

$$Mg(s) + H_2SO_4(aq) \longrightarrow MgSO_4(aq) + H_2(g)$$

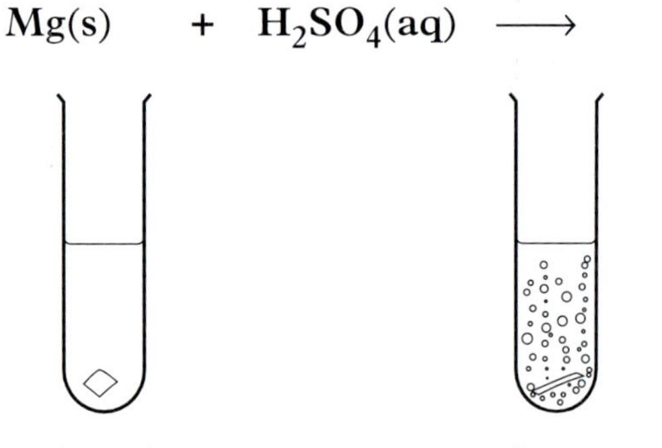

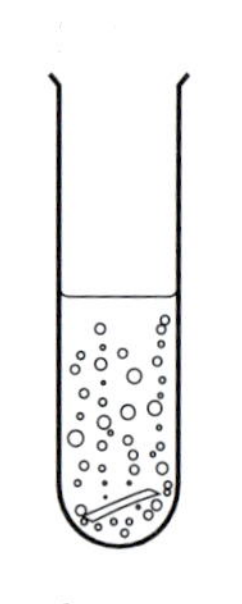

copper doesn't react

magnesium reacts vigorously with acids, releasing hydrogen

calcium reacts violently with acids, releasing hydrogen

You can see from these reactions that there is a pattern

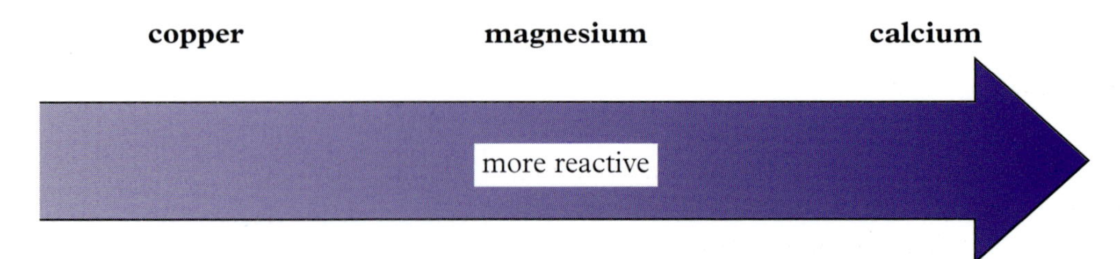

copper
magnesium
calcium
more reactive

Questions

1 Which property or properties of a metal would make it useful as the following?
   a) a frying pan
   b) electrical connecting wires
   c) horse shoes.

2 What is meant by the term **malleable**?

3 Name three magnetic materials.

4 What is formed when metals burn in air?

5 Name two metals which dissolve in cold water.

6 Which gas is formed when metals react with cold water?

7 Describe what happens when a reactive metal is placed in dilute acid.

8 List the following metals in order of their reactivity. Start with the least reactive metal.

   calcium copper magnesium

The reactivity list of metals

All metals can be listed in order of their reactivity. Here is a list of just some of the metals:

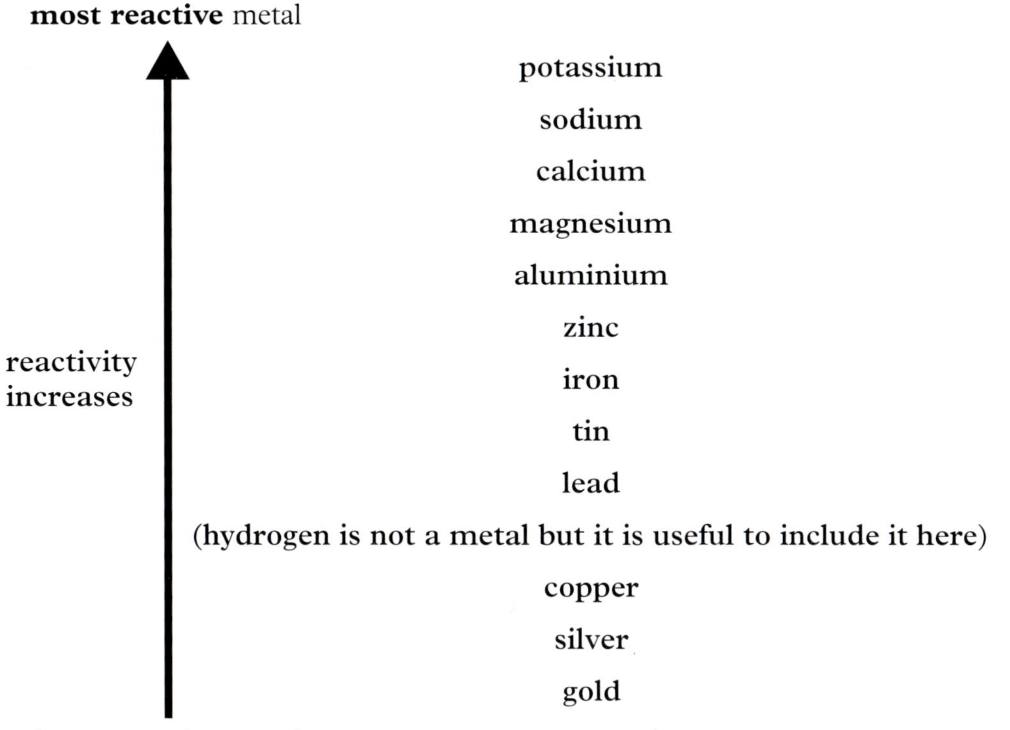

most reactive metal
potassium
sodium
calcium
magnesium
aluminium
zinc
reactivity increases
iron
tin
lead
(hydrogen is not a metal but it is useful to include it here)
copper
silver
gold
least reactive metal

It is easy to see that magnesium is more reactive than copper just by burning each metal, but many metals are more alike in their reactivity. Another way of comparing metals is to look at **displacement reactions**.

Displacement reactions of metals

If you add one metal to a compound of a different metal, the metal which is the **more reactive** of the two will end up in the compound. The more reactive metal will **displace** the less reactive metal from the compound.

For example, you can add a metal to another metal oxide and heat them:

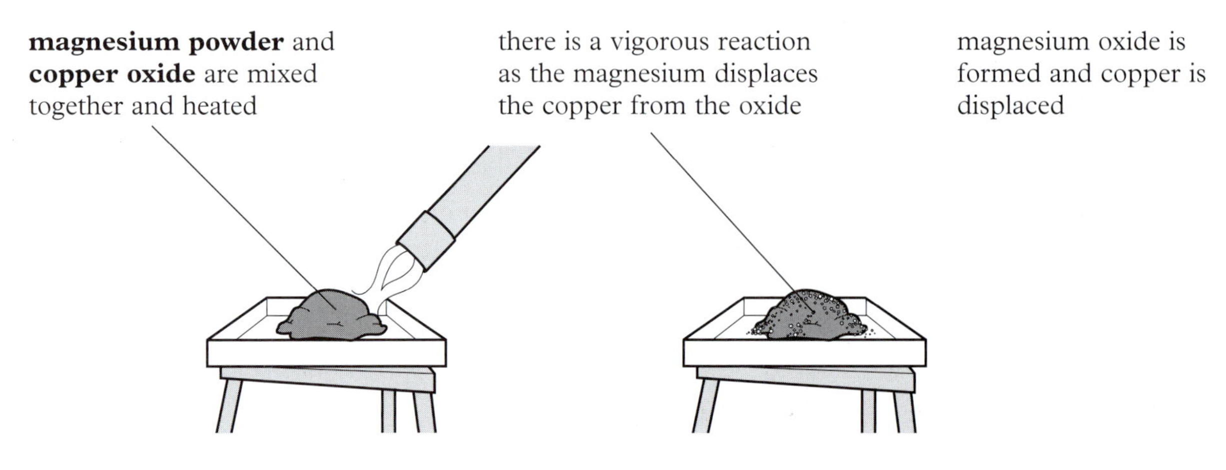
magnesium powder and **copper oxide** are mixed together and heated
there is a vigorous reaction as the magnesium displaces the copper from the oxide
magnesium oxide is formed and copper is displaced

The following equation explains what happened in the experiment above:

$$\textbf{magnesium + copper oxide} \xrightarrow{\text{heat}} \textbf{copper + magnesium oxide}$$

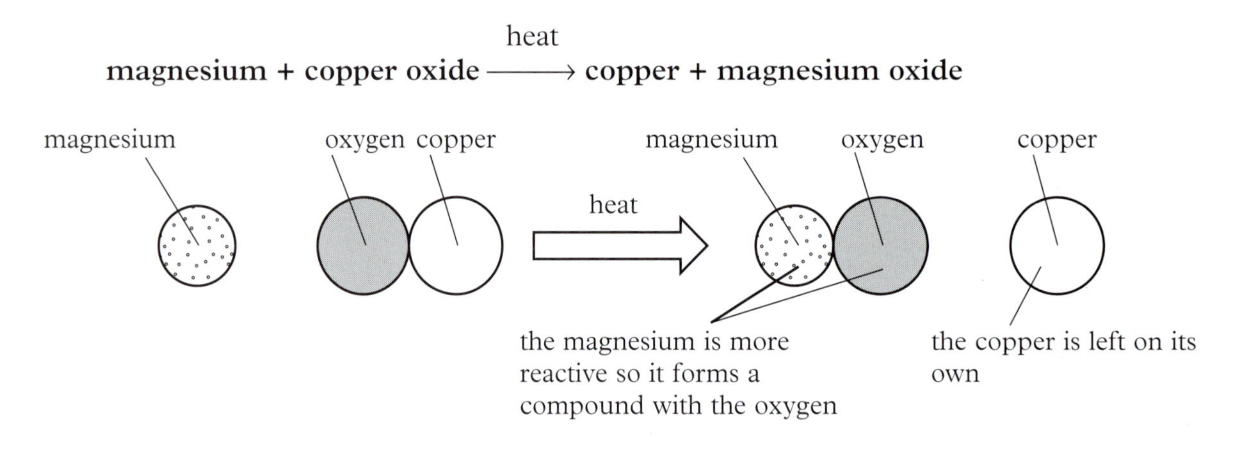
magnesium
oxygen copper
heat
magnesium
oxygen
copper
the magnesium is more reactive so it forms a compound with the oxygen
the copper is left on its own

If we now heat magnesium oxide and copper together there is no reaction. The magnesium keeps the oxygen, because it is more reactive than copper.

Displacement reactions in a test-tube

You can add one metal to another metal compound which is dissolved in water. The more reactive metal will displace the less reactive one as before.

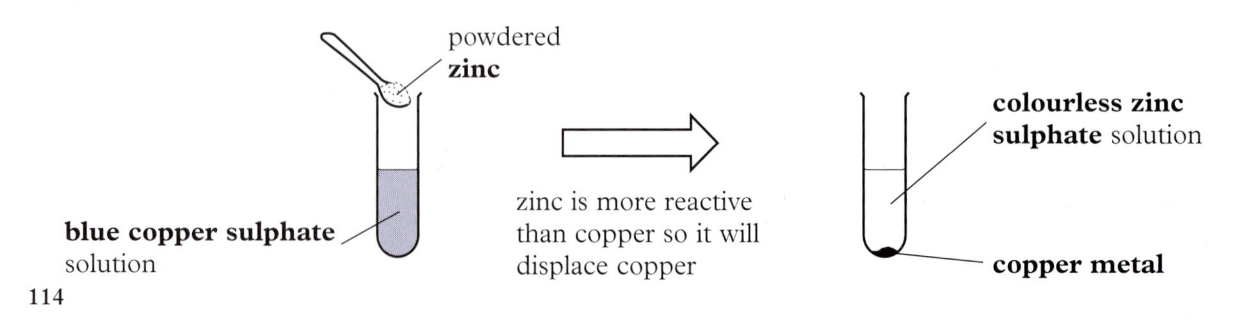
powdered **zinc**
blue copper sulphate solution
zinc is more reactive than copper so it will displace copper
colourless zinc sulphate solution
copper metal

This is the reaction:

zinc + copper sulphate ⟶ zinc sulphate + copper

$$Zn(s) + CuSO_4(aq) \longrightarrow ZnSO_4(aq) + Cu(s)$$

The general rule is that the more reactive metal ends up in the compound. You can tell which is the more reactive metal by looking at the reactivity list on page 113. The more reactive metal is the one which is higher up the list.

The strange case of aluminium

Aluminium is a reactive metal quite high up in the reactivity list. You might think it would react with the air and gradually corrode away. Yet, you probably have saucepans at home which are made of aluminium, and you might wrap your sandwiches in aluminium foil. Neither of these things corrode away.

Aluminium metal is always covered with a very very thin layer of alumunium oxide, which sticks to the surface like a coat of paint. This layer of aluminium oxide protects the metal from further reactions with the air or water.

Using the reactivity list

The reactivity list will help you to predict:

- How metals will react with water and acids.
- Whether metal A will displace metal B from a compound.

  If metal A is above metal B in the list, metal A will displace metal B from its compounds.

  The further apart the metals are in the list, the more vigorous the reaction.

Questions

1 Copy and complete the table below. Use the reactivity list on page 113 to decide whether there would be a reaction between the metal and the compound.

2 Would a ring made of gold react if you put it into a beaker of dilute acid? Explain your answer.

metal	compound	reaction takes place? yes/no	word equation which describes reaction
magnesium	zinc oxide		
zinc	iron oxide		
copper	calcium oxide		
iron	sodium chloride		
gold	copper sulphate		
zinc	magnesium sulphate		

The corrosion of metals

Nearly every metal will react with chemicals around it, such as oxygen in the air, water, or acids. Each metal does so at a different rate. Some metals will react so slowly that it seems as if they are not reacting at all. The more reactive the metal, the faster it will react.

For example, sodium is a metal that visibly reacts with air and water very quickly. However, iron rusts slowly in the presence of air and water. Gold doesn't react at all.

The rusting of iron

Iron is a cheap, strong metal that has many many uses. Iron rusts or corrodes in the presence of air and water. The rusting of iron costs countries millions of pounds, which are spent replacing rusted iron or protecting new iron.

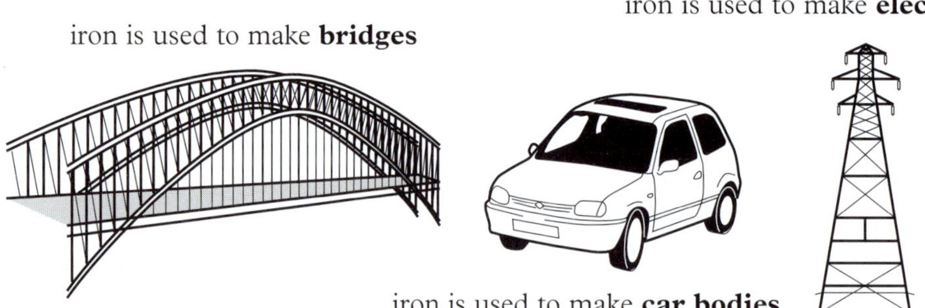
iron is used to make electricity pylons
iron is used to make bridges
iron is used to make car bodies

Iron rusts away to iron oxide in the presence of air and water. A simple experiment can be set up to show when rusting will take place. The experiment starts with **shiny iron nails** which are put into three different test-tubes and **left for a week**.

After a week the test-tubes look like the ones below:

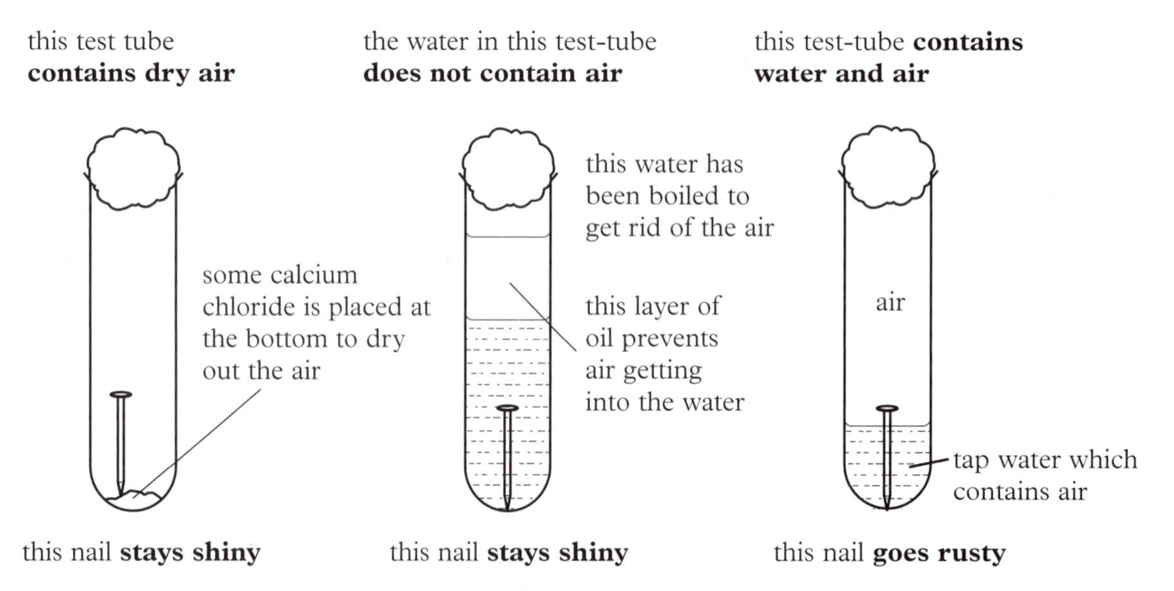
this test tube contains dry air
the water in this test-tube does not contain air
this test-tube contains water and air
this water has been boiled to get rid of the air
some calcium chloride is placed at the bottom to dry out the air
this layer of oil prevents air getting into the water
air
tap water which contains air
this nail stays shiny
this nail stays shiny
this nail goes rusty

How to prevent rusting

If air or water are stopped from getting to the iron, then the iron does not rust. Chromium plating prevents air and water getting to the iron and also looks good.

Sacrificial protection

A more reactive metal is attached to the iron. The more reactive metal reacts instead and this protects the iron. This method of preventing rusting is called **sacrificial protection**.

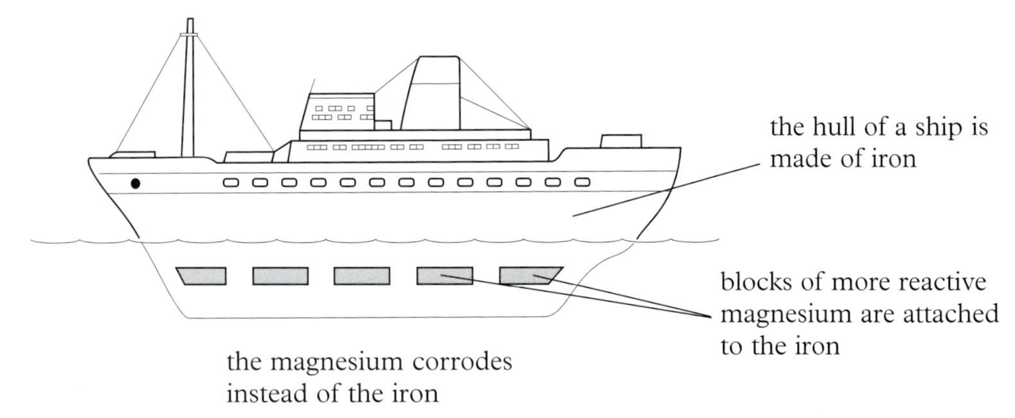
the hull of a ship is made of iron
blocks of more reactive magnesium are attached to the iron
the magnesium corrodes instead of the iron

Galvanising

Iron can be coated with zinc to prevent rusting. The zinc coating stops the water and air from getting to the iron, and also protects the iron because zinc is more reactive than iron. Coating iron with zinc is called **galvanising**. Grey coloured metal dustbins are galvanised.

Questions

1 Why does the rusting of iron cost millions of pounds?

2 Describe an experiment you could carry out to show the conditions needed for rusting.

3 Describe what is meant by:
a) chromium plating
b) sacrificial protection
c) galvanising.

4 Ships are made of steel which is a form of iron. Shipwrecks often stay in surprisingly good condition at the bottom of the ocean. Explain why a shipwreck can stay in good condition whilst it is surrounded by water.

Alloys

Metals can be made more useful by mixing them with small amounts of other metals and sometimes with non-metals. The substance produced is called an **alloy**.

Alloy	Mixture	Special properties	Examples of uses
Brass	Copper and zinc	Doesn't corrode, hard, attractive	Wind instruments like trumpets
Duralumin	Aluminium, copper, manganese, magnesium	Low density, strong	Aircraft bodies
Solder	Tin and lead	Easily melted	Joining metal surfaces
Nichrome	Nickel and chromium	Doesn't corrode, high electrical resistance	Wire in electrical heating elements

Summary

Metals react with oxygen to form metal oxides. If the metal oxides dissolve in water the solutions formed are alkaline. Only the most reactive metals react with water and release hydrogen, while moderately reactive metals react with acids and release hydrogen.

The reactivity list puts the metals in order of reactivity with the most reactive metal at the bottom. A more-reactive metal will displace a less-reactive one from a compound. Metals corrode at different rates – the more reactive the metal, the faster it corrodes. The corrosion of iron is called rusting. Iron is a widely used metal and it is prevented from rusting by stopping air and water from coming into contact with it. A more-reactive metal may also be attached to the iron. This metal then reacts and stops the iron from rusting. Alloys are mixtures of metals which have properties that make them more useful than the pure metals.

Key words

corrosion The reaction of a metal with other substances such as oxygen so that the metal is changed into a metal compound, like the metal oxide.

displacement reactions Reactions between a metal and a metal compound. The more reactive metal displaces the less reactive metal. The general rule is that the more reactive metal, which is the one higher up the reactivity list, ends up in the compound.

metals Hard and shiny solids which are good conductors of heat and electricity.

reactivity How rapidly and vigorously a chemical change takes place. A very reactive material reacts faster than a less reactive one to make a new substance.

End of Chapter 11 questions

1 One of the properties of metals is that they are malleable – you can hammer them into new shapes without them breaking.

a) Pick three other properties that are typical of metals from the following list:

conduct heat well **shiny** **brittle**

are poor conductors of electricity **magnetic** **solid**

b) The table below gives information about a metal, its use and its properties. Copy and complete the table with the details of three other metals. Put the most suitable use and properties from the list in the correct place in the table.

Metal	Use	Properties
Nickel	Coins	Doesn't corrode and is shiny
Aluminium		
Copper		
Tin		

very good conductor of heat and electricity **coating steel food cans**

doesn't corrode and is not poisonous **saucepans**

doesn't corrode and low density **electrical wires**

2 The table shows some of the reactions of four metals – A, B, C, and D

Metal	Reaction with dilute hydrochloric acid	Reaction with water
A	Vigorous, gives off a gas	Slow, gives off a few bubbles of gas
B	Slow, gives off a gas	No visible reaction
C	No reaction	No reaction
D	Not done	Vigorous, gives off a gas

a) Place the metals in order of reactivity, with the most reactive first.

b) The gas given off could be hydrogen. Describe a test for hydrogen and the result you would expect (see page 184).

c) The reaction of D with dilute hydrochloric acid was not tried. Explain why.

d) Which metal could be

i) copper

ii) calcium?

e) Zinc, Zn, reacts with dilute hydrochloric acid, HCl, to form zinc chloride, $ZnCl_2$ and hydrogen, H_2.

i) Write the word equation for this reaction

ii) This is the balanced symbol equation:

$$Zn(s) + 2HCl(aq) \longrightarrow ZnCl_2(aq) + H_2(g)$$

What do the letters in brackets mean? (See page 67.)

f) Describe how you would do this experiment.

3 Iron window frames corrode after a few years but ones made of aluminium do not, even if they are not painted.

a) What word is used to describe the corrosion of iron?

b) What two substances in the air cause iron to corrode?

c) How may iron window frames be protected from corrosion other than by being painted?

d) Aluminium is a more reactive metal than iron. Explain why unpainted aluminium frames do not corrode, while iron ones do.

4 Here is part of the reactivity list of metals

sodium

magnesium

aluminium

——

iron

lead

copper

gold

a) Which of the following metals could be placed in the gap?

silver potassium zinc lithium

b) When magnesium and copper oxide are mixed and heated, there is a vigorous reaction, producing magnesium oxide and copper.

i) Explain why this reaction takes place readily but there is no reaction between copper and magnesium oxide.

ii) In which of the following mixtures would you expect a reaction when heated?:

- magnesium and lead oxide
- lead and magnesium oxide
- gold and iron oxide.

5 The lists give some properties of iron and of sulphur.

Iron	**Sulphur**
Property	**Property**
Shiny	Low melting point
Magnetic	Brittle
Conducts electricity	Does not conduct electricity
High melting point	Solid

a) Give two properties from the list which show that iron is a metal.

b) Give two properties from the list which show that sulphur is a non-metal.

c) Some of the properties of aluminium metal are given below.

conducts electricity well **resists corrosion** **conducts heat well**

low density for a metal **has a shiny surface**

Aluminium has a wide variety of uses. For each of the uses below, give one of the properties from the list above that makes aluminium suitable for:

i) making saucepans
ii) making aircraft bodies
iii) making electricity power lines
iv) making window frames.

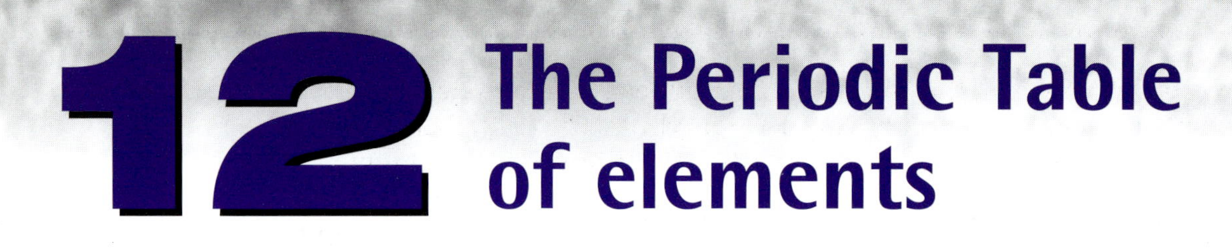

12 The Periodic Table of elements

The elements

It took a long time for people to puzzle out what the world was made of. From the time of the Ancient Greeks, people thought that there were just four elements – fire, air, earth and water.

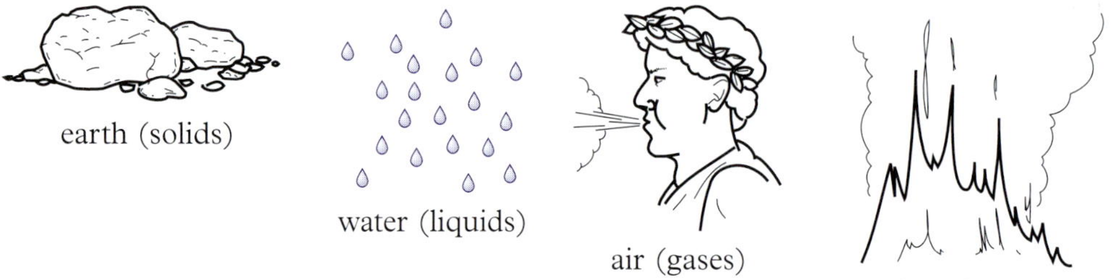
earth (solids)
water (liquids)
air (gases)
fire (flames)

For nearly a thousand years it was thought that all solids were 'earth' elements. This made people think that lead and gold were made from the same basic material. People called alchemists actually tried to change lead into gold.

The elements

About three hundred years ago it was thought that there were some materials that could not be broken down into anything else. These materials were called **elements**. We know now that there are 109 elements and more will probably be found.

Sorting out the elements

Although there are 109 different elements, they can be sorted out into similar sets. The first division is into metals and non-metals. After that, metals or non-metals that are alike can be grouped together.

The Periodic Table

A table that contains all the elements, set out in such a way that elements that are like each other appear together, is called the **Periodic Table**.

The early versions of the Periodic Table listed the elements in order of their relative atomic mass, A_r (see page 33). The relative atomic mass is the average mass of an atom compared with the mass of an atom of hydrogen.

Today, the **elements are listed in order of atomic number**, **Z**, starting with hydrogen, which has an atomic number of 1.

The **atomic number** of an element **is the number of protons in the nucleus**. An atom of helium has **two protons** in its nucleus and therefore has **atomic number 2**.

Here is the Periodic Table. The table is split into rows called **Periods** which are numbered **1–7** and columns called **Groups** which are numbered **O–VII**. You need to understand the key to this table. Each element is represented by a box. Each box contains information about one element. The first diagram explains how sodium is shown on the Periodic Table. Sodium is in Group I, Period 3.

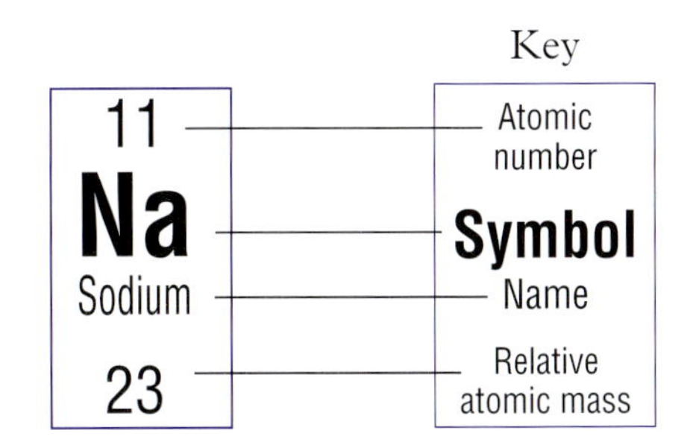
Key
11
Na
Sodium
23
Atomic number
Symbol
Name
Relative atomic mass

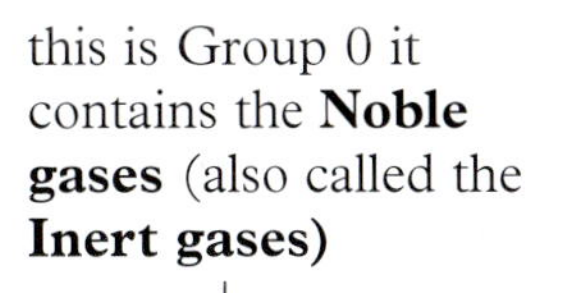
this is Group 0 it contains the **Noble gases** (also called the **Inert gases)**

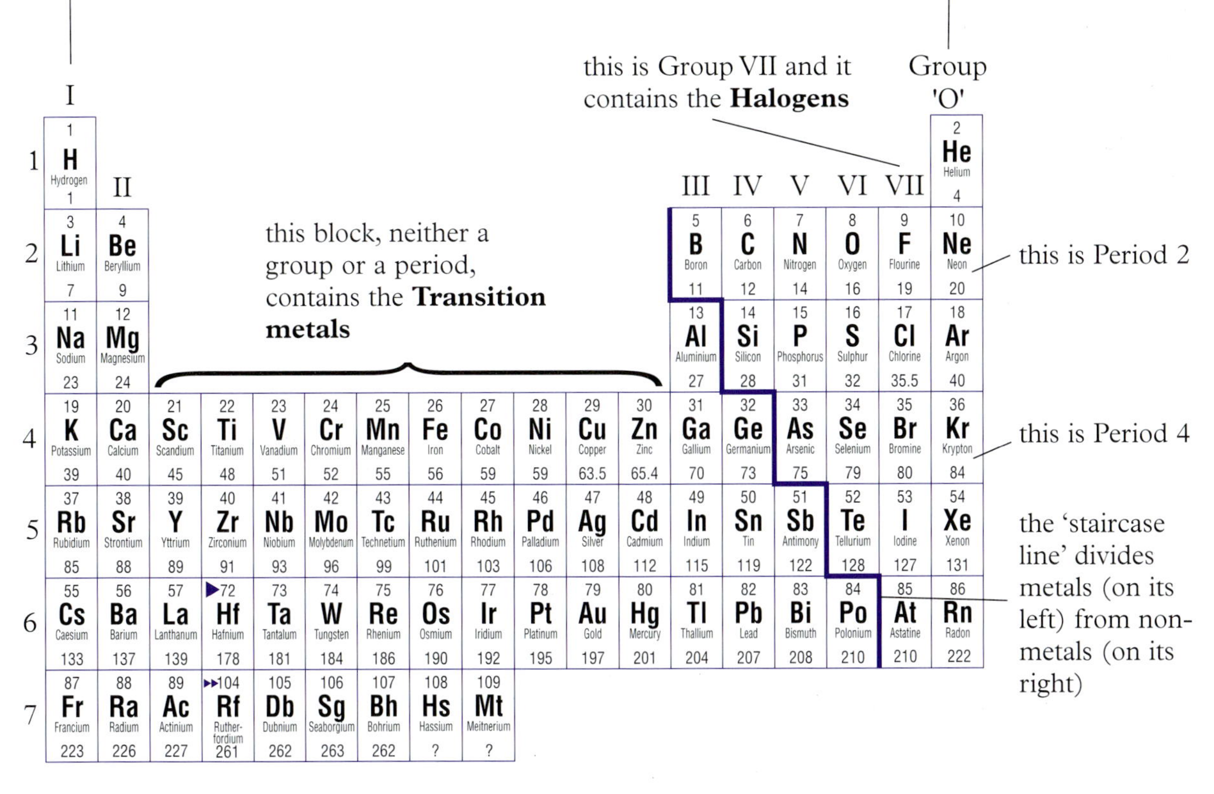
this is Group I and it contains the **Alkali metals**
this is Group VII and it contains the **Halogens**
Group 'O'
this block, neither a group or a period, contains the **Transition metals**
this is Period 2
this is Period 4
the 'staircase line' divides metals (on its left) from non-metals (on its right)

Period	I	II											III	IV	V	VI	VII	0
1	1 H Hydrogen 1																	2 He Helium 4
2	3 Li Lithium 7	4 Be Beryllium 9											5 B Boron 11	6 C Carbon 12	7 N Nitrogen 14	8 O Oxygen 16	9 F Flourine 19	10 Ne Neon 20
3	11 Na Sodium 23	12 Mg Magnesium 24											13 Al Aluminium 27	14 Si Silicon 28	15 P Phosphorus 31	16 S Sulphur 32	17 Cl Chlorine 35.5	18 Ar Argon 40
4	19 K Potassium 39	20 Ca Calcium 40	21 Sc Scandium 45	22 Ti Titanium 48	23 V Vanadium 51	24 Cr Chromium 52	25 Mn Manganese 55	26 Fe Iron 56	27 Co Cobalt 59	28 Ni Nickel 59	29 Cu Copper 63.5	30 Zn Zinc 65.4	31 Ga Gallium 70	32 Ge Germanium 73	33 As Arsenic 75	34 Se Selenium 79	35 Br Bromine 80	36 Kr Krypton 84
5	37 Rb Rubidium 85	38 Sr Strontium 88	39 Y Yttrium 89	40 Zr Zirconium 91	41 Nb Niobium 93	42 Mo Molybdenum 96	43 Tc Technetium 99	44 Ru Ruthenium 101	45 Rh Rhodium 103	46 Pd Palladium 106	47 Ag Silver 108	48 Cd Cadmium 112	49 In Indium 115	50 Sn Tin 119	51 Sb Antimony 122	52 Te Tellurium 128	53 I Iodine 127	54 Xe Xenon 131
6	55 Cs Caesium 133	56 Ba Barium 137	57 La Lanthanum 139	▶72 Hf Hafnium 178	73 Ta Tantalum 181	74 W Tungsten 184	75 Re Rhenium 186	76 Os Osmium 190	77 Ir Iridium 192	78 Pt Platinum 195	79 Au Gold 197	80 Hg Mercury 201	81 Tl Thallium 204	82 Pb Lead 207	83 Bi Bismuth 208	84 Po Polonium 210	85 At Astatine 210	86 Rn Radon 222
7	87 Fr Francium 223	88 Ra Radium 226	89 Ac Actinium 227	▸▸104 Rf Rutherfordium 261	105 Db Dubnium 262	106 Sg Seaborgium 263	107 Bh Bohrium 262	108 Hs Hassium ?	109 Mt Meitnerium ?									

this block of elements fits into the main table in Periods 6 and 7, shown by the arrows

▶ 58 Ce Cerium 140	59 Pr Praseo-dymium 141	60 Nd Neo-dymium 144	61 Pm Promethium 147	62 Sm Samarium 150	63 Eu Europium 152	64 Gd Gadolinium 157	65 Tb Terbium 159	66 Dy Dysprosium 163	67 Ho Holmium 165	68 Er Erbium 167	69 Tm Thulium 169	70 Yb Ytterbium 173	71 Lu Lutetium 175
▸▸ 90 Th Thorium 232	91 Pa Protact-inium 231	92 U Uranium 238	93 Np Neptunium 237	94 Pu Plutonium 242	95 Am Amercium 243	96 Cm Curium 247	97 Bk Berkelium 245	98 Cf Californium 251	99 Es Einsteinium 254	100 Fm Fermium 253	101 Md Mendel-evium 256	102 No Nobelium 254	103 Lr Lawren-cium 257

Questions

1 Give the symbol of the element in:
   a) Group I, Period 3
   b) Group II, Period 3
   c) Group 0, Period 4.

2 Give the Group and Period of:
   a) sodium, Na
   b) chlorine, Cl
   c) neon, Ne.

3 Give the symbol for:
   a) the inert gas with the smallest atomic number
   b) four transition metals beginning with C
   c) the halogen beginning with C
   d) the alkali metal with the largest atomic number.

4 Draw a table with two columns. Head one column **metals** and the other **non-metals**. Write the name and symbol of each element below in the appropriate column.

   phosphorus, P; indium, In; iodine, I; cerium, Ce; astatine, At; krypton, Kr; gadolinium, Gd.

Groups in the Periodic Table

The elements in a group in the Periodic Table are all very similar. Elements in the same group have the same number of electrons in their outer shell. The number of the group tells you how many electrons are in the outer shell.

Although the elements in each group are similar, there are small differences between each one. As you read down the group the properties and reactivities of the series gradually change.

Group I: The alkali metals

Although hydrogen is often put at the top of Group I, it is not at all typical. Hydrogen is a gas and all the other elements in the group are metals. We call these the **alkali metals**.

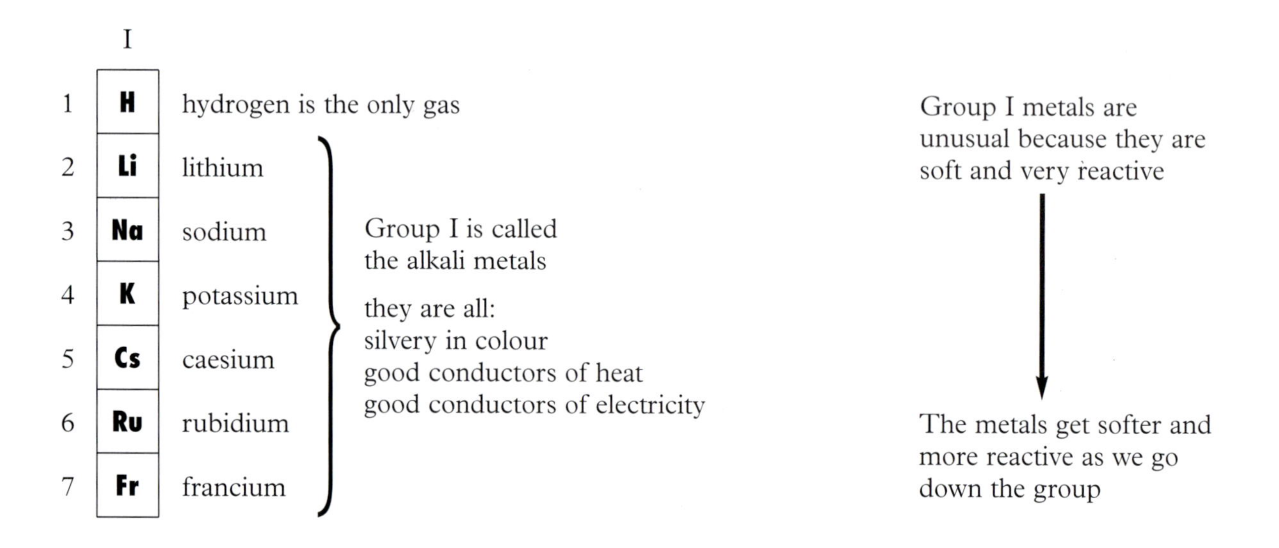
I
1 H hydrogen is the only gas
2 Li lithium
3 Na sodium
4 K potassium
5 Cs caesium
6 Ru rubidium
7 Fr francium
Group I is called the alkali metals
they are all:
silvery in colour
good conductors of heat
good conductors of electricity
Group I metals are unusual because they are soft and very reactive
The metals get softer and more reactive as we go down the group

The atomic structures of the first three metal elements in Group I

All group I elements have **full inner shells** and **one electron** in their **outer shell**.

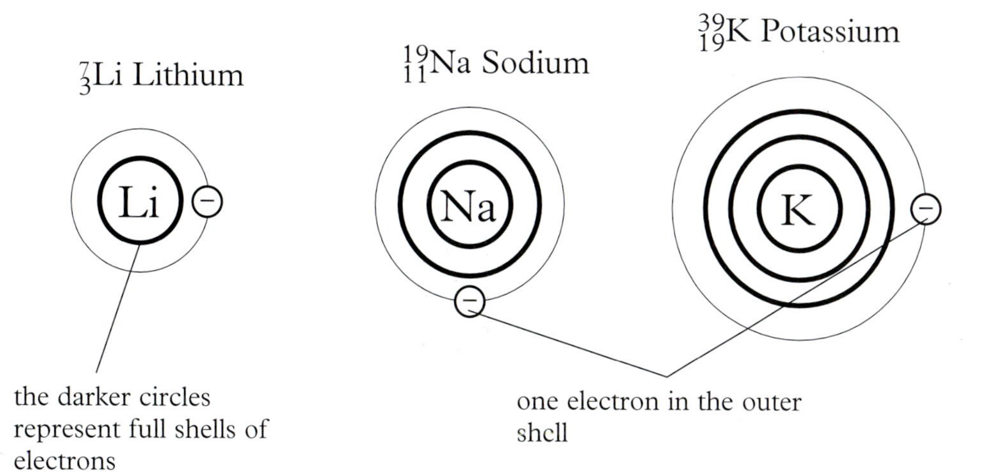

$^{7}_{3}$Li Lithium
$^{19}_{11}$Na Sodium
$^{39}_{19}$K Potassium
Li
Na
K
the darker circles represent full shells of electrons
one electron in the outer shell

The properties of the alkali metals

Appearance: The metals are silver when freshly cut. Otherwise, they are covered with a coating of the metal oxide and look grey and dull.

The **metals are soft**: lithium is a dark silver metal which can be cut with some pressure. Sodium is a silver metal which cuts as easily as hard cheese. Potassium is a silver metal that cuts as easily as soft cheese.

Other properties of these three metals are shown in the table below:

Metal	Density/g/cm^3	Melting temperature/°C	Boiling temperature/°C
Lithium	0.53	181	1342
Sodium	0.97	98	883
Potassium	0.86	63	760

Questions

1 The density of water is 1 g/cm^3. From the data in the table above, would the alkali metals float or sink in water?

2 Look carefully at the table above. What is the trend in how easily the metals melt as you go down the group?

The reactions of the alkali metals with air

All the alkali metals are stored in oil to help keep the air away. The alkali metals react quickly with the oxygen in air to form a layer of metal oxide. The reaction with air gets faster with the more reactive metals.

Group I metals are called the **alkali metals** because their oxides are strongly alkaline.

Alkali metals burn brightly in air to form metal oxides. These metal oxides dissolve in water to form very alkaline solutions. Universal Indicator paper turns purple with a pH of about 12 in such a solution.

The reactions of the alkali metals with water

The alkali metals all react with water. During this reaction hydrogen is released. The metal oxide which is formed dissolves in the water. In this way an alkaline solution is left at the end of the reaction.

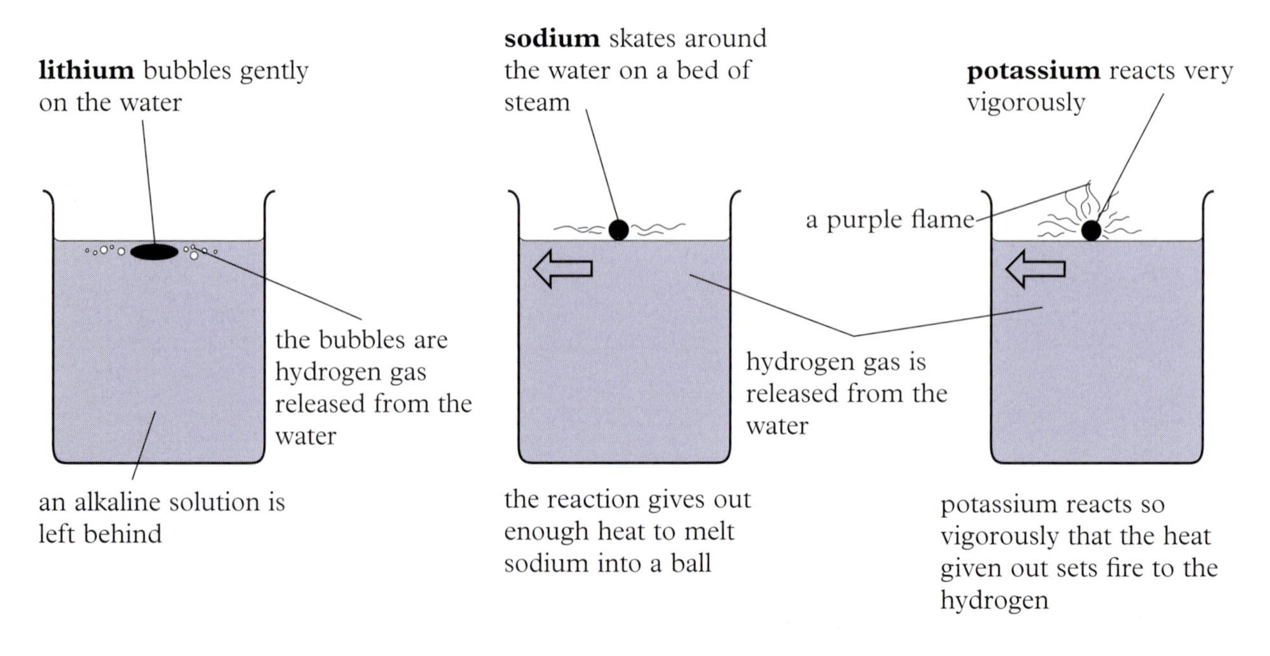
lithium bubbles gently on the water
sodium skates around the water on a bed of steam
potassium reacts very vigorously
a purple flame
the bubbles are hydrogen gas released from the water
hydrogen gas is released from the water
an alkaline solution is left behind
the reaction gives out enough heat to melt sodium into a ball
potassium reacts so vigorously that the heat given out sets fire to the hydrogen

Questions

1 Explain why lithium, sodium and potassium are called alkali metals.

2 Potassium is a very soft grey solid. What property of potassium will tell you that it is a metal?

3 The next metal below potassium in Group I is rubidium.
   a) What would you expect the metal to look like?
   b) How would you expect it to react with water? Explain your answer.

The compounds of alkali metals

The Group I metals themselves are not very useful because they are so reactive, but the compounds they form with non-metals are the opposite. They are **unreactive** and they have many uses. These compounds are white, and all dissolve in water. Sodium compounds have many uses:

Compound	Uses
sodium carbonate Na_2CO_3	• Soda glass used for milk bottles is made from sodium carbonate and sand. • Washing soda is sodium carbonate which is used to make hard water soft. Washing soda is often added to washing powders.
sodium chloride NaCl	• Sodium chloride is processed to make table salt. • Rock salt is spread onto icy roads. It has the effect of lowering the freezing temperature of ice, so that the ice melts.
sodium hydroxide NaOH	• Sodium hydroxide is used to make soaps and detergents. • Bleaches are made from sodium hydroxide and chlorine.
sodium hydrogencarbonate $NaHCO_3$	• Antacids for indigestion tablets may contain sodium hydrogencarbonate to neutralise acids in the stomach. • Sodium hydrogencarbonate in baking powder releases carbon dioxide with acids so that doughs will rise.

Group VII The Halogens

The halogens are elements on the right hand side of the Periodic Table. All the halogens are reactive non-metals. The elements get less and less reactive as we go down the group.

Group VII

Period	Symbol	Name
2	**F**	fluorine
3	**Cl**	chlorine
4	**Br**	bromine
5	**I**	iodine
6	**At**	astatine

Group VII is called the **halogens**

all the halogens have a bleachy smell, like the smell in a swimming pool

most reactive halogen

↓

least reactive halogen

The atomic structure of the first three elements in Group VII

All Group VII elements have **full inner shells** and **seven electrons** in their outer shell.

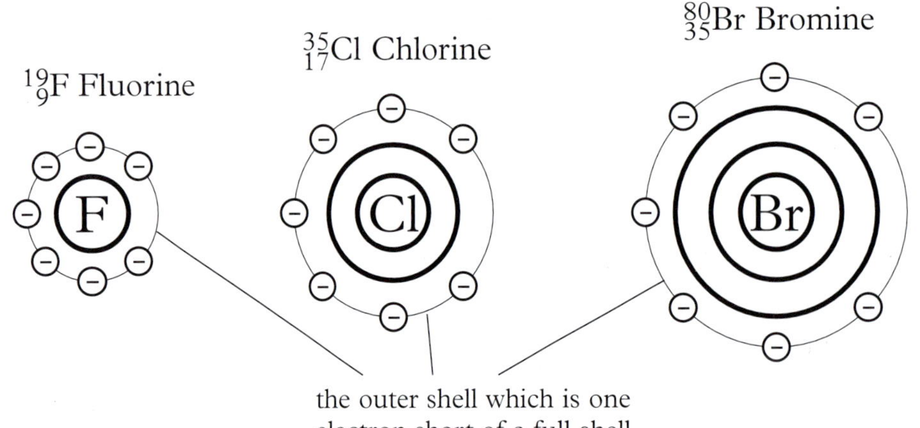

$^{19}_{9}$F Fluorine
$^{35}_{17}$Cl Chlorine
$^{80}_{35}$Br Bromine
F
Cl
Br
the outer shell which is one electron short of a full shell

As elements chlorine, fluorine and bromine exist as molecules made up of pairs of atoms

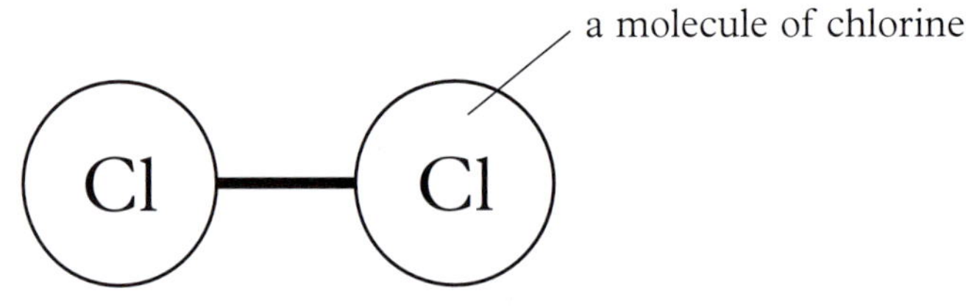

a molecule of chlorine
Cl
Cl

The properties of the halogens

Fluorine is such a reactive gas that it is dangerous and difficult to handle. It is a pale yellow gas. We will look in more detail at the next three members of the group:

Element	Appearance	Melting temperature / °C	Boiling temperature / °C
Chlorine	yellow-green gas	−101	−35
Bromine	brown liquid	−7	59
Iodine	shiny grey-black solid	144	184

The reactions of the halogens to form bleaching solutions

All the halogens have a bleachy smell, which is like the smell of a swimming pool.

Element	Reaction with water	Reaction with sodium hydroxide solution
Chlorine	Dissolves to form a bleaching solution	Dissolves to form a strongly bleaching solution. This solution is in the bleach that you buy for use at home.
Bromine	Dissolves to form a weakly bleaching solution	Dissolves to form a bleaching solution
Iodine	Does not dissolve	Dissolves to form a weakly bleaching solution

The reaction of halogens with metals

The halogens combine directly with metals to form salts called metal halides with no other products.

metal + halogen → salt

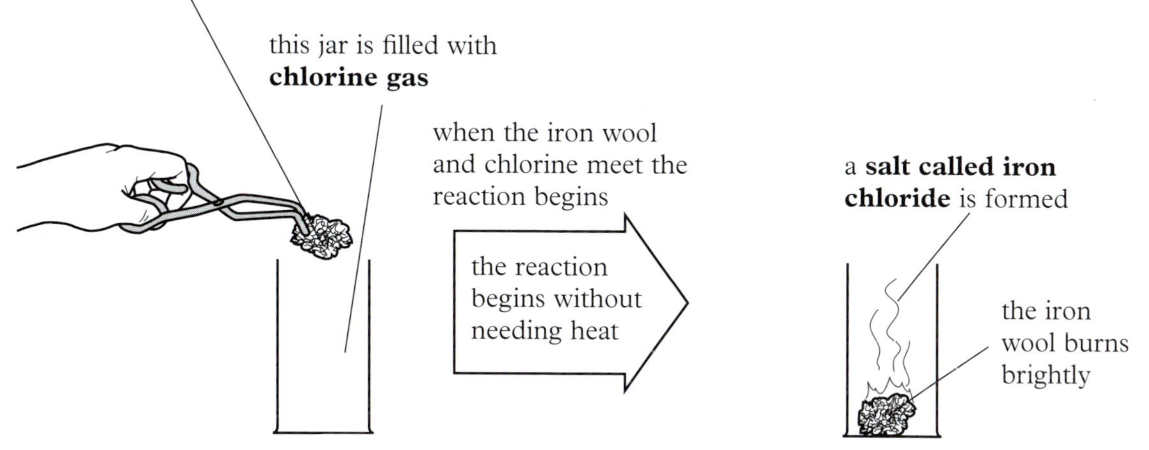
iron wool is a metal
this jar is filled with **chlorine gas**
when the iron wool and chlorine meet the reaction begins
the reaction begins without needing heat
a **salt called iron chloride** is formed
the iron wool burns brightly

iron + chlorine → iron chloride

The reaction of iron with bromine

The reaction of iron in a gas jar of bromine vapour needs heat to start and is much less vigorous. The reaction is:

iron + bromine → iron bromide

The reaction of iron with iodine

The reaction with iodine only works if the iron wool is heated strongly.

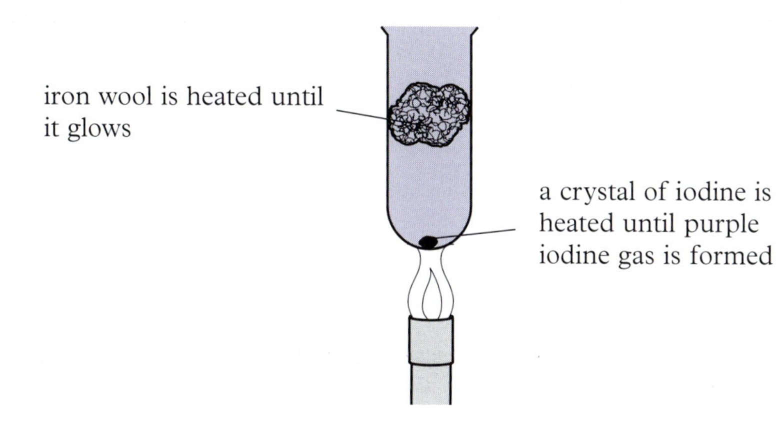
iron wool is heated until it glows
a crystal of iodine is heated until purple iodine gas is formed

iron + iodine → iron iodide

Again, this shows how the halogens get less reactive as we go down the group.

Displacement reactions

The more reactive the element, the more likely it is to combine with other elements. In this sort of reaction a compound is formed. If a halogen element is mixed with a solution of a compound containing a less reactive halogen element, the more reactive halogen ends up in the compound. The more reactive halogen **displaces** the less reactive halogen in the compound.

A displacement reaction takes place when chlorine gas is bubbled through a solution of potassium bromide.

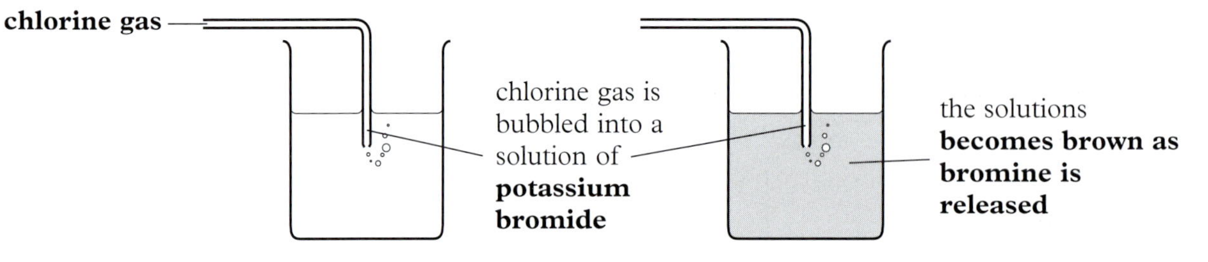
chlorine gas
chlorine gas is bubbled into a solution of **potassium bromide**
the solutions **becomes brown as bromine is released**

The equation for this reaction is:

chlorine + potassium bromide → bromine + potassium chloride

Questions

1 Are the compounds of alkali metals reactive or unreactive?

2 a) What is another name for the elements in Group VII of the Periodic Table?
b) How many electrons do the Group VII elements have in their outer shells?
c) Complete the following reaction:
metal + halogen → ________

3 Copy and complete the sentence below using words from the following list:
more less

A displacement reaction takes place when a ________ reactive halogen displaces a ________ reactive halogen.

Uses of the halogens and their compounds

Chlorine forms a bleach in water which kills germs. Chlorine is used to purify our drinking water and the water in swimming pools.

Iodine solution is used as an antiseptic for cuts and grazes.

Compounds of the halogens

Salts of the halogens are called **halides**.

The **silver halides**, for example silver bromide and silver iodide, darken when they are put in the light. Both silver bromide and silver iodide are used in photography.

The **hydrogen halides** such as hydrogen chloride are gases. In water the hydrogen halides form strong acids. For example hydrogen chloride gas is very soluble in water. It forms hydrochloric acid.

hydrogen chloride $\xrightarrow{\text{water}}$ **hydrochloric acid**

$$HCl(g) \xrightarrow{\text{water}} HCl(aq)$$

Hydrochloric acid has many uses in industry. The acid is used for cleaning metals, making medicines and making other useful chemicals.

Whenever chemists dilute a concentrated acid, they say this rhyme: '*Always remember that you ought-er add the acid to the water.*'

You should never add water to the acid. This is because the reaction gives out a lot of heat and if you add water to a concentrated acid, the water could boil and spit and injure you.

The transition metals

The transition metals are mostly **hard and not very reactive metals**. This means that they are very useful to us in their normal state.

In the Periodic Table clusters of transition metals that are near to each other are very similar.

Copper, **Cu**, and **nickel**, **Ni**, are used to make **coins**.
Iron, **Fe**, **cobalt**, **Co**, and **nickel**, **Ni**, are **magnetic**.
Silver, **Ag**, **platinum**, **Pt**, and **gold**, **Au**, are precious metals and are used in **jewellery**.

Many of the transition metal compounds are coloured. For example, copper sulphate is a bright blue.

The metals are often **used as catalysts**. For example, iron is used as the catalyst in making ammonia.

Group 0 The Noble gases (also called the Inert gases)

Gases from Group 0 are found in tiny proportions in the air we breathe.

The noble gases are very unreactive gases. They are the only elements that exist as single separate atoms. As we go down the group, the density and the boiling temperature of each gas increases.

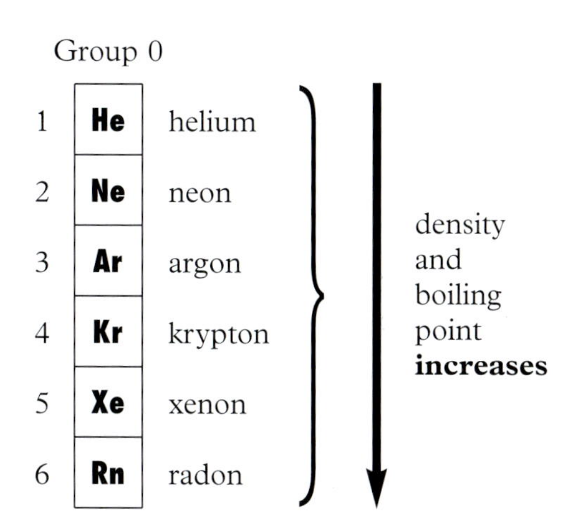
Group 0
1
He
helium
2
Ne
neon
3
Ar
argon
4
Kr
krypton
5
Xe
xenon
6
Rn
radon
density
and
boiling
point
increases

All the noble gases have a full outer shell of electrons. Three of the noble gases are shown below:

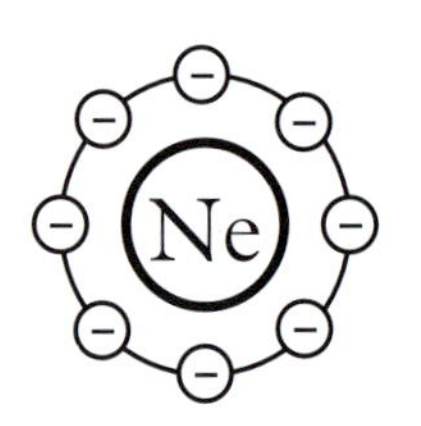
Ne

neon has **two full shells** of electrons

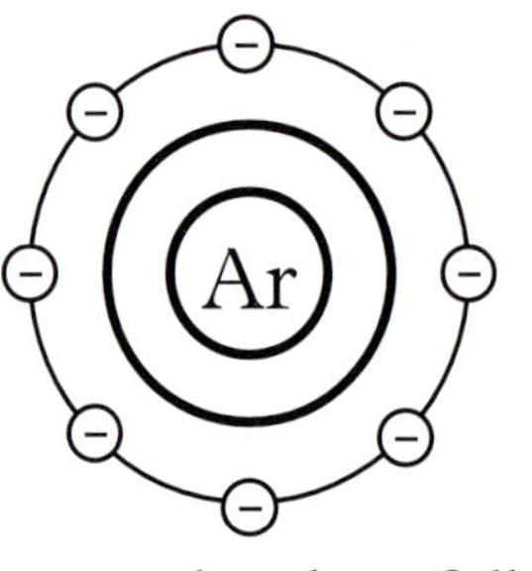
Ar

argon has **three full shells** of electrons

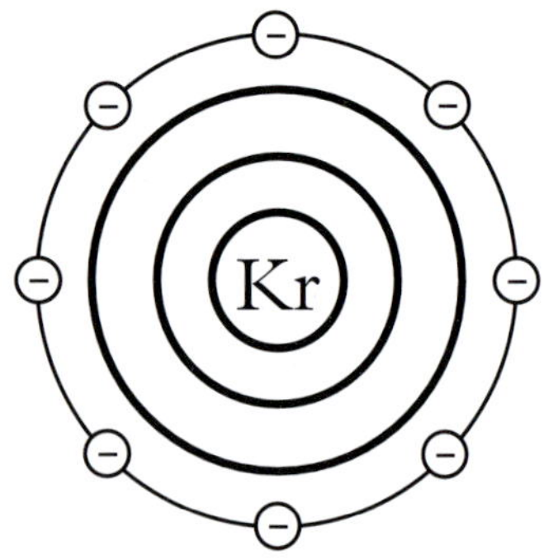
Kr

krypton has **four full shells** of electrons

Some of the uses of the Noble gases

Noble gas	Use
Helium	As a gas to lift airships and keep them in the air Mixed with oxygen in breathing apparatus for deep-sea divers
Neon	In fluorescent light tubes
Argon	In electric light bulbs Blown around high temperature welding to stop the metal reacting with oxygen

Summary

The Periodic Table of elements contains all the elements listed in order of atomic number. The table is arranged so that elements that are alike appear in columns. These columns are called Groups. There are eight Groups. Periods go from left to right. On the left hand side of the table are the metals and on the right side are the non-metals.

Group I contains the alkali metals. They are reactive metals. Alkali metals get softer and also **more** reactive as we go down the group. The oxides of alkali metals are alkaline.

Group VII contains the halogens. These are reactive non-metals. They have a bleachy smell. Halogens get **less** reactive as we go down the group.

Group 0 contains the noble or inert gases. These are very unreactive gases which exist as single atoms.

The transition metals are a block of metals in the middle of the table. They are hard, unreactive metals with many uses.

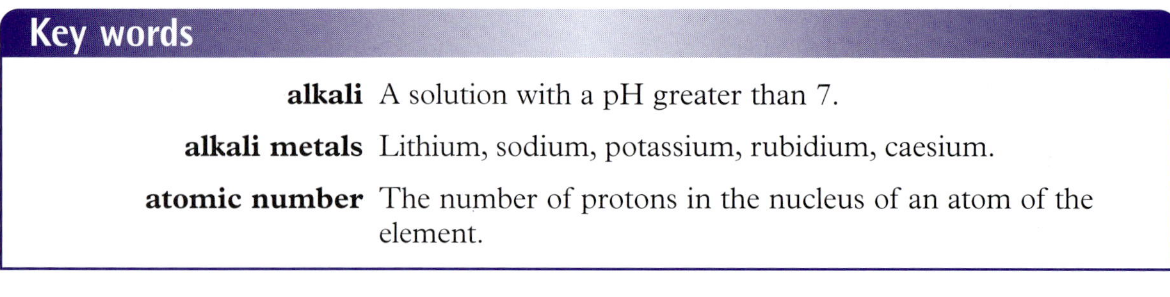

Key words

alkali A solution with a pH greater than 7.

alkali metals Lithium, sodium, potassium, rubidium, caesium.

atomic number The number of protons in the nucleus of an atom of the element.

Key words (continued)

halogens Fluorine, chlorine, bromine, iodine, astatine.

relative atomic mass, A_r This is the mass of one atom of an element compared with the mass of one atom of hydrogen.

End of Chapter 12 questions

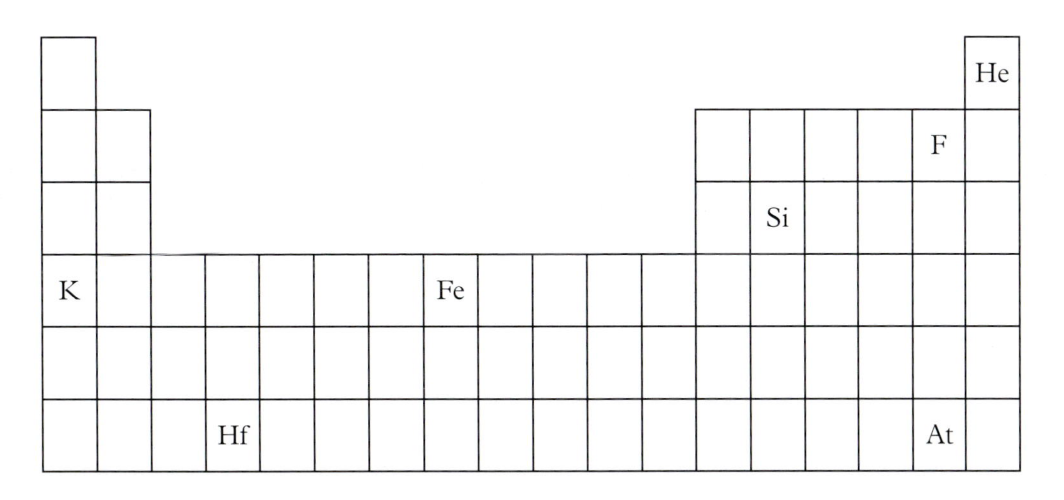
He
F
Si
K
Fe
Hf
At

1 Use only the elements marked in the above outline of the Periodic Table, and pick:
   a) an element in Group I
   b) an element from Period 3
   c) a noble gas
   d) a transition metal
   e) the most reactive element from Group VII.

2 Use the Periodic Table above to answer this question. Is Hf, hafnium, a metal or a non-metal? Explain your answer.

3 Sodium is a reactive metal in Group I of the Periodic Table. It is a fairly soft, shiny metal that floats on water. It reacts with water, producing sodium hydroxide and hydrogen.
   a) Write a word equation for the reaction of sodium with water.
   b) Describe what you would see if you placed a small lump of sodium in a large beaker of water.
   c) Sodium hydroxide is an alkali. How could you show that an alkali was produced in this reaction?

4 Lithium is above sodium in the Periodic Table.
   a) How would the reaction of lithium with water be similar to the reaction of sodium with water?
   b) How would the reaction of lithium with water be different from the reaction of sodium with water?
   c) Sodium and lithium are normally stored in jars of paraffin oil. Give two reasons why this is done.

5 The table below shows some properties of three of the halogen elements listed in the order in which they appear in the Periodic Table.

Element	Colour	State at room temperature
Chlorine		gas
Bromine	red-brown	
Iodine		solid

a) Copy the table and fill in the blanks using any of the words from the list below:

blue gas solid green black liquid

b) Fluorine is above chlorine in the Periodic Table. Predict the state of fluorine at room temperature.

c) Give one chemical reaction that all the halogens have in common and write a word equation for it.

d) Say how the reactivity of the halogens changes as we go down the group in the Periodic Table.

e) i) Give one use for chlorine.

ii) Silver bromide is used in photography. Explain why this is.

6

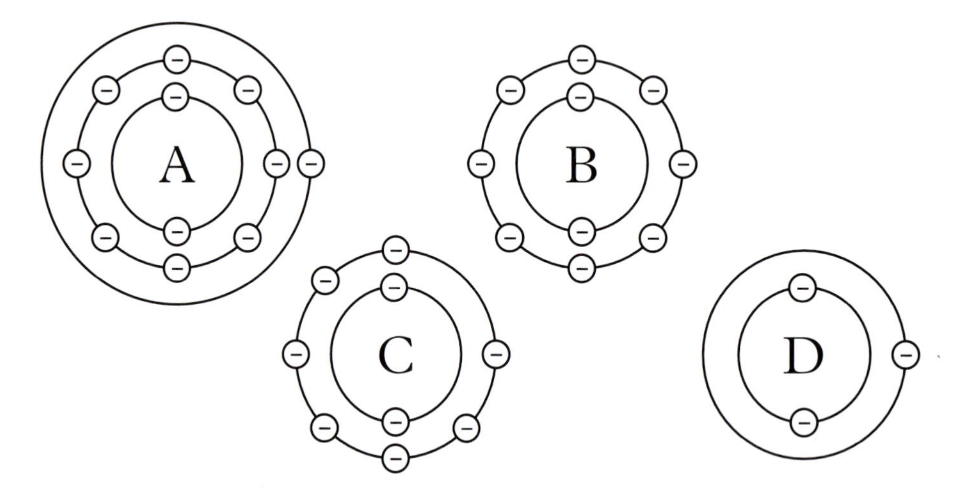
A
B
C
D

Look at the electron arrangements of some elements shown above. These are not their real symbols.

a) (i) Give the letter of an element that exists as separate atoms.

(ii) Explain why this element does not react.

b) (i) Give the letters of two elements in Group I of the Periodic Table. Explain why you have chosen these two elements.

(ii) Which of these elements will be more reactive? Explain your answer.

c) (i) Give the letter of the element in Group VII of the Periodic Table.

(ii) When this element in Group VII reacts with a metal it gains a full shell of electrons and becomes an ion. What is the charge on the ion?

d) Using the letters as the symbols, give the formula of a compound formed between one of the elements in Group I and one of the elements in Group VII. Explain your answer.

13 Useful products from air

Air is a mixture of about 80% nitrogen, N_2, and 20% oxygen, O_2. There are also very small percentages of carbon dioxide, CO_2, water vapour and all the inert gases, including helium, neon and argon. Nitrogen and oxygen are both used in industry. Each is used on its own or to make other chemicals. Even the inert gases are useful to us when separated from the air. Air is a very cheap **raw material** or a starting material for making chemicals.

How the gases in air are separated

All the gases in air may be separated by **fractional distillation**. First, air passes through a filter to remove dust. Then it is turned into a liquid by cooling it to around −200 °C under pressure. Next it is distilled. The liquid air is allowed to warm up. Nitrogen boils off first, leaving oxygen as a pale blue liquid. The noble gases may also be separated out in the process, because they boil off at different temperatures.

Pure oxygen gas is used in steel making, metal cutting and for medical purposes. Oxygen is also used in chemical processes without being separated from the air. Fuels like coal and oil burn in the oxygen in air to give out heat. The oxygen in the process for making sulphuric acid, H_2SO_4, is supplied by the air.

Nitrogen

Nitrogen is a very unreactive gas, so most of the time it stays in the air not reacting with anything. Despite this it is a very important element.

Plants need nitrogen to build the proteins found in their leaves and stems. Most plants cannot take nitrogen directly from the air. They take nitrogen up from the soil in the form of chemicals called **nitrates**.

How nitrogen from the air is taken up by plants

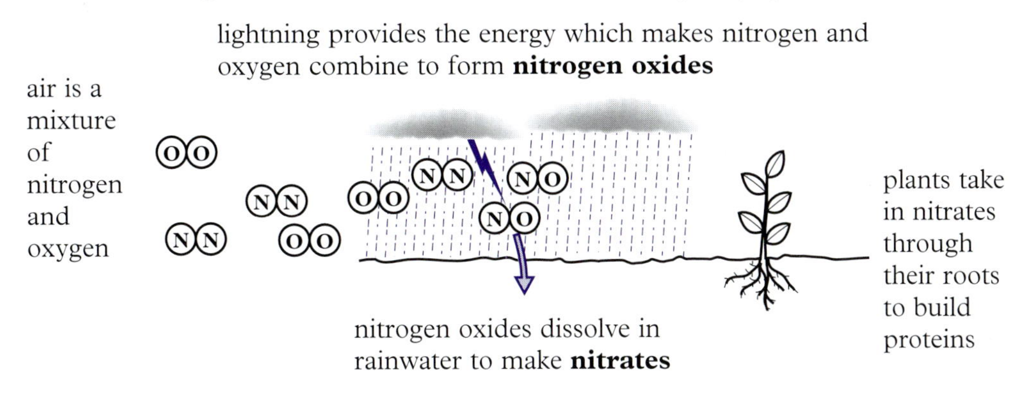
lightning provides the energy which makes nitrogen and oxygen combine to form **nitrogen oxides**
air is a mixture of nitrogen and oxygen
O O
N N
N O
plants take in nitrates through their roots to build proteins
nitrogen oxides dissolve in rainwater to make **nitrates**

Making fertilisers from nitrogen

The world's population is getting larger all the time. We need to grow as many crops as possible in order to feed everyone. These crops need more nitrogen than nature can supply. We add fertilisers to supply nitrogen and other chemicals to the land. Fertilisers are made using the nitrogen in air as a raw material. Three stages are needed to change nitrogen from the air into a fertiliser:

1 Nitrogen is reacted with hydrogen to make a gas called **ammonia**, by the **Haber Process**.

2 Ammonia is reacted with oxygen to make **nitric acid**.

3 Ammonia and nitric acid react together to make the fertiliser **ammonium nitrate**.

Making ammonia

This is the ammonia molecule.

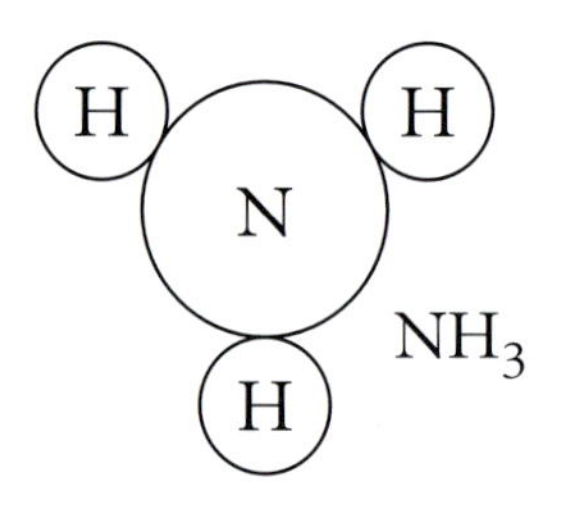
H
H
N
H
NH_3

Ammonia is a gas with a characteristic smell.

Ammonia dissolves in water to make an alkaline solution called **ammonium hydroxide**.

The Haber Process for making ammonia

The basic reaction used to make ammonia is:

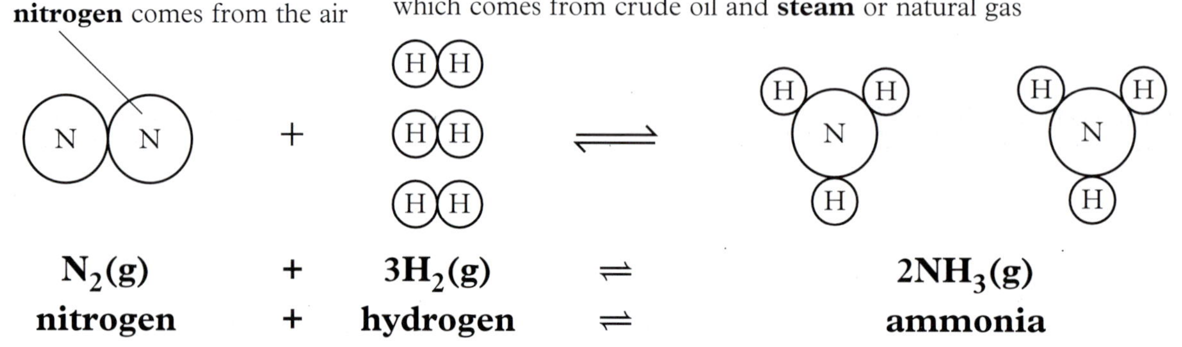
hydrogen is produced from a reaction between **methane** which comes from crude oil and **steam** or natural gas
nitrogen comes from the air
N N
+
H H
H H
H H
H N H H
H N H H

$$N_2(g) + 3H_2(g) \rightleftharpoons 2NH_3(g)$$

nitrogen + hydrogen ⇌ ammonia

The sign ⇌ means that the reaction is reversible. As soon as nitrogen and hydrogen react to form ammonia, some of the ammonia turns back into nitrogen and hydrogen. So we end up with a **mixture of nitrogen, hydrogen and ammonia**. This is a problem, because we are trying to make ammonia.

By carefully choosing the right temperature and pressure, a useful amount of ammonia can be produced. An iron catalyst is used to speed up the reaction. The best conditions for the process are about 450 °C and a pressure which is 200 times greater than atmospheric pressure.

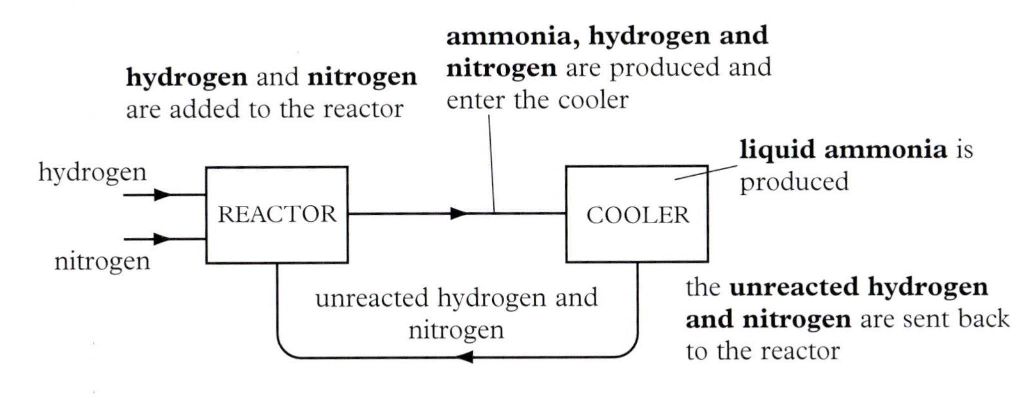

ammonia, hydrogen and nitrogen are produced and enter the cooler
hydrogen and nitrogen are added to the reactor
liquid ammonia is produced
hydrogen
nitrogen
REACTOR
COOLER
unreacted hydrogen and nitrogen
the unreacted hydrogen and nitrogen are sent back to the reactor

Making nitric acid from ammonia

Ammonia is used to make nitric acid, $\mathbf{HNO_3}$. Nitric acid is used to make **fertilisers** and **explosives**. Making nitric acid from ammonia involves two stages:

1 **Ammonia** is first **reacted with oxygen** from the air. Nitrogen monoxide and water are produced. A platinum catalyst is used.

2 Nitrogen monoxide reacts with more oxygen and water to give **nitric acid**.

Overall the reaction is:

ammonia + oxygen ⟶ nitric acid + water

Making the fertiliser ammonium nitrate from nitric acid

Ammonia is an alkali. When ammonia is dissolved in water it reacts with acids to make salts and water. The solid salts may then be used as fertilisers.

Nitric acid and ammonium hydroxide react together to produce the salt ammonium nitrate and water.

nitric acid + ammonium hydroxide ⟶ ammonium nitrate + water

Questions

1 What proportion of the air is nitrogen?
2 Explain how nitrogen from the air is changed into a substance which plants can use.
3 Why do we add fertilisers to the land?
4 Describe the three main stages used to change nitrogen from the air into the fertiliser ammonium nitrate.
5 In the Haber Process for making ammonia:
a) What does the sign $\rightleftharpoons$ mean?
b) Why is this a problem in the production of ammonia?
6 a) What elements are present in ammonia?
b) Which one of these elements is important for fertilisers?

Questions (continued)

7 a) What is a raw material?
b) From which raw material does the nitrogen in ammonia come?
c) From which two raw materials does the hydrogen in ammonia come?

8 Ammonium nitrate has the formula NH_4NO_3. Why is it a useful fertiliser?

9 Why do we need to produce fertilisers?

Problems with fertilisers

Ammonium nitrate is a widely used fertiliser. This fertiliser is very soluble in water. Ammonium nitrate can cause problems in rivers and lakes.

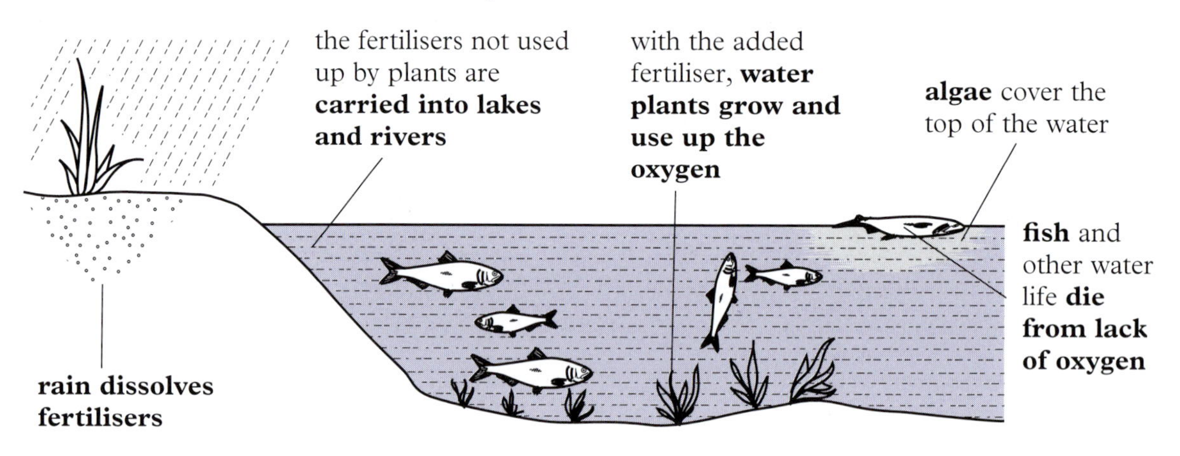
the fertilisers not used up by plants are **carried into lakes and rivers**
with the added fertiliser, **water plants grow and use up the oxygen**
algae cover the top of the water
fish and other water life **die from lack of oxygen**
rain dissolves fertilisers

Fertilisers in drinking water

Nitrates are difficult to remove from water. Small amounts of nitrates may be present in your drinking water if you live in an area where there is a lot of crop farming. There is some evidence that suggests nitrates in water may cause health problems, especially in babies.

Summary

The two main gases in air are nitrogen (about 80%) and oxygen (about 20%). These may be separated by fractional distillation.

In the Haber Process, nitrogen and hydrogen react together to form ammonia:

$$N_2(g) + 3H_2(g) \rightleftharpoons 2NH_3(g)$$

nitrogen + hydrogen ⇌ ammonia

This is a reversible reaction.

Ammonia and oxygen react together to form nitric acid. Ammonia is reacted with nitric acid to produce ammonium nitrate. Ammonium nitrate is used as a fertiliser which helps crops grow. Fertilisers can cause problems if they are present in lakes, rivers and drinking water.

Key words

acids Solutions with pHs of less than 7.

alkalis Solutions with pHs of more than 7.

fertilisers Compounds that supply nitrogen and other elements for growing plants.

fractional distillation A separation process in which a mixture of liquids is heated and the fractions which boil off over different temperature ranges are collected.

salts Substances formed when an acid and an alkali react together:

acid + alkali ⟶ salt + water

End of Chapter 13 questions

1 Air is a mixture of about 80% nitrogen with several other gases.

a) What is the second most abundant gas in the atmosphere? Describe a test for this gas and the result that you would expect. (See page 184.)

b) Nitrogen is needed by plants in order for them to grow. However, most plants cannot use nitrogen directly from air, it must first be made into soluble nitrogen compounds.

(i) Describe how nitrogen in the air is made into soluble nitrogen compounds in nature.

(ii) Give the name of the process by which nitrogen in the air is combined with hydrogen to form ammonia (NH_3).

(iii) Complete the balanced symbol equation for this process

$$N_2 + \_\_\_ H_2 \rightleftharpoons \_\_\_ NH_3$$

(iv) What does the symbol $\rightleftharpoons$ mean?

(v) Where does the hydrogen for this process come from?

c) Ammonia is an alkaline gas. To make it easier to apply to fields, it is converted into the solid salt ammonium sulphate.

(i) Why is a solid salt easier to apply than a gas?

(ii) Which acid would be needed to make ammonium sulphate from ammonia?

2 Fill in the blanks in the passage below using words from the following list. Each word may be used once, more than once or not at all.

increase	**reversible**	**soluble**	**ammonium nitrate**
nitrogen	**hydrogen**	**neutralised**	
oxygen	**temperature**	**insoluble**	

The element n_______ is an unreactive gas found in air which is vital for plant growth. In the Haber Process, this element is made to combine with h_______ to make ammonia, NH_3. The reaction which is used is r_______ , so the conditions of t_______ and pressure have to be carefully chosen to make sure that we get a good yield of ammonia. To make fertilisers, ammonia is n_______ with nitric acid to make the salt a_______ . However, this salt is s_______ in water and can run off into rivers and streams where it can i_______ the growth of water plants.

3 Air is a mixture of gases. Some of the gases are listed below:

oxygen carbon dioxide nitrogen argon

a) Which of these gases makes up the biggest part of the air?
b) Two of the gases in the list combine together during lightning flashes. Name these two gases.
c) One of the gases in the list is made up of two elements.
   (i) Name this gas and the two elements in it.
   (ii) How does the name of this gas tell you that it contains only two elements?
d) (i) Which of the gases in the list is an inert gas?
   (ii) What does the term inert mean? (See page 131.)
   (iii) Suggest a use of this gas which makes use of its inertness.
   (iv) Name another inert gas found in the atmosphere which is not in the list above.

4 Read the following passage and answer the questions that follow:

All the gases in air may be separated by fractional distillation. The air is turned into a liquid by cooling it to around –200 °C under pressure. The liquid air is allowed to warm up. Nitrogen boils off first, leaving oxygen as a blue liquid. The noble gases are also separated out in the process. Oxygen gas is used in deep-sea diving and in hospitals. Nitrogen is used to fill crisp packets and to surround steel when it is being heat treated.

a) What is distillation?
b) Why is this called **fractional** distillation?
c) Which gas has the higher boiling temperature – oxygen or nitrogen?
d) Why is oxygen useful in hospitals and deep-sea diving?
e) Why is nitrogen used to fill crisp packets and in heating steel?

14 Useful products from rocks

Some rocks, such as limestone, slate or granite, are quarried and then the rock itself is used for building. Other rocks are quarried or mined and then processed to produce substances like iron and aluminium. Metals like iron and aluminium are found in the rocks around us. The metals are in compounds which make up rocks. These compounds are called **minerals**. Iron is found in a metal compound called iron oxide. Aluminium is extracted from the compound aluminium oxide.

the **compounds** that make up rocks are called **minerals**
some rocks are a **mixture of more than one mineral**
haematite is a form of **iron oxide**
quartzite is silicon dioxide

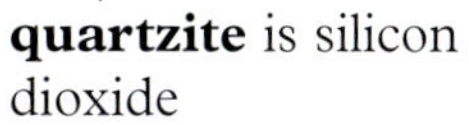

In a **metal compound** the metal is chemically bonded to another element or elements.

A **metal ore** is a rock which is worth mining to extract the metal from its compound. A mining company works out whether it can make enough money from the metal before the rocks are mined.

Sometimes metal ores are not mined because there are too many problems to overcome. The metal ore may be too near to where people live. Mining may not be allowed in an area of special scientific interest or natural beauty. Mining is not worthwhile if there is not enough of the mineral that contains the metal. In some areas the metal ore may be so difficult to reach that it would cost too much to mine and extract it.

Questions

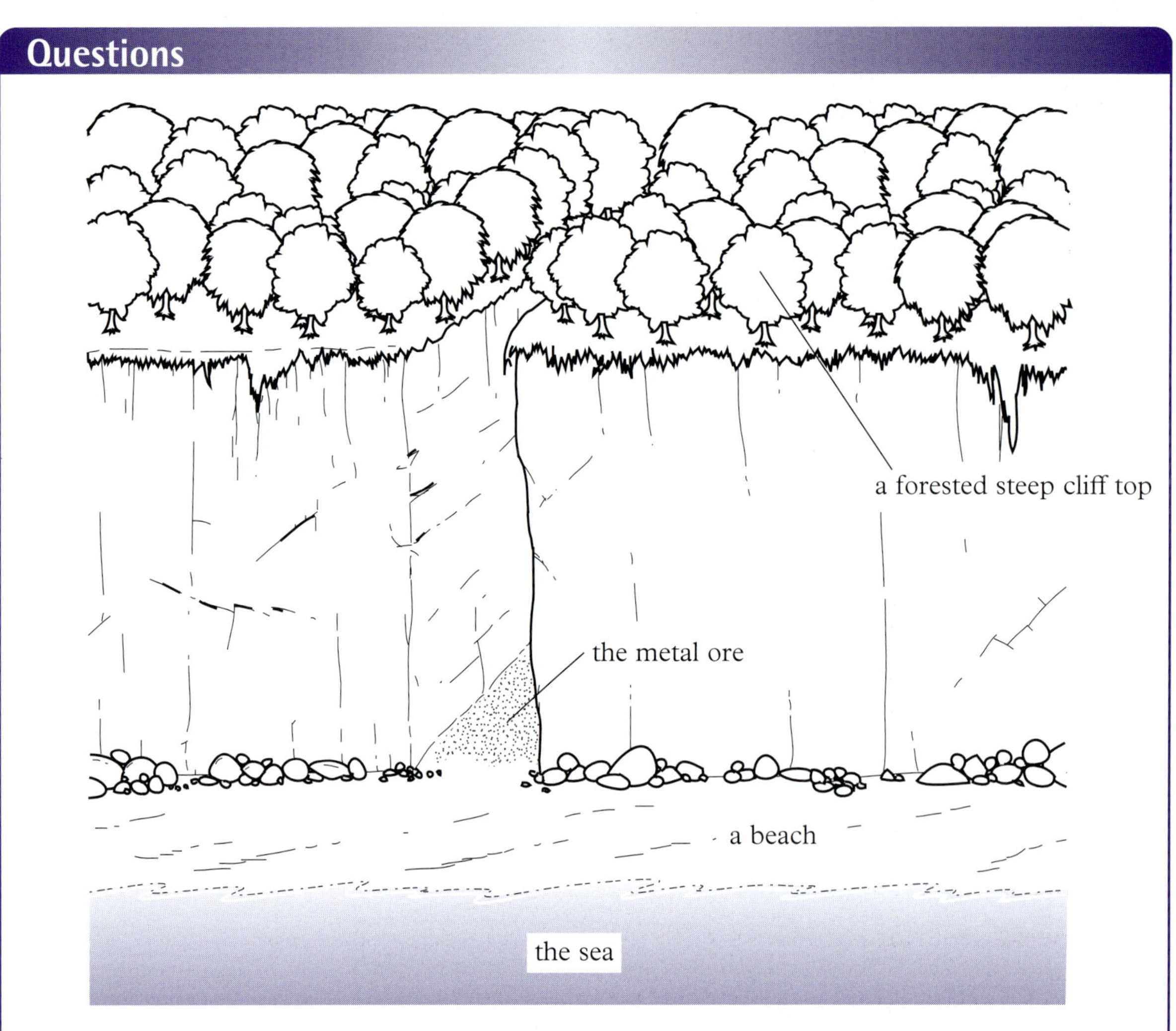
a forested steep cliff top
the metal ore
a beach
the sea

1 Look at the diagram above. You are in charge of mining for the metal ore.

a) Make a list of all the problems there are.
b) Write down ways that you might get at the ore.
c) Make a list of all the items you would need to pay for to extract the ore.
d) What might make you decide to go ahead?

2 Working with a partner, make a drawing of another scene where it would be hard (but not impossible) to mine the metal ore. Swap your drawing with that of another pair and try and solve the problems involved in mining the ore in their scene.

Separating the metal from its compound

After the ore is mined, the mineral containing the metal compound is separated from the other substances in the rocks. Then the metal is freed from its compound. How this is done depends on how reactive the metal is.

We can list all the metals in order of how reactive they are. Here is a list of the more common ones. Carbon is not a metal but is included in the list to show you where it comes in the order of reactivity.

To help us work out how to obtain the free metal, the metals are divided into three main groups: unreactive metals, fairly reactive metals and very reactive metals.

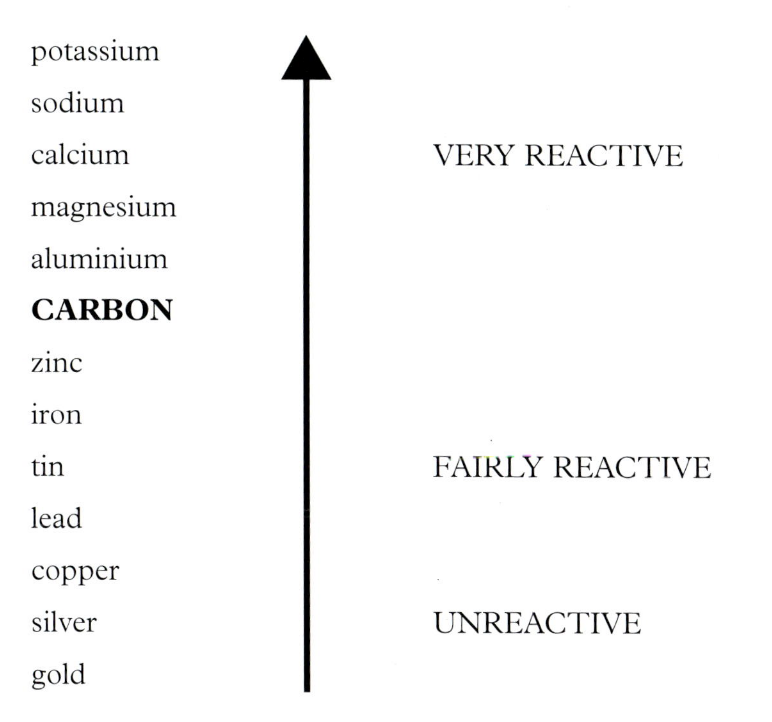
potassium
sodium
calcium
magnesium
aluminium
CARBON
zinc
iron
tin
lead
copper
silver
gold
VERY REACTIVE
FAIRLY REACTIVE
UNREACTIVE

Unreactive metals

These **metals are not combined with other elements** and so can be mined directly. The rock containing the metal is crushed. The metal is denser than the rock fragments and this makes it possible to separate them.

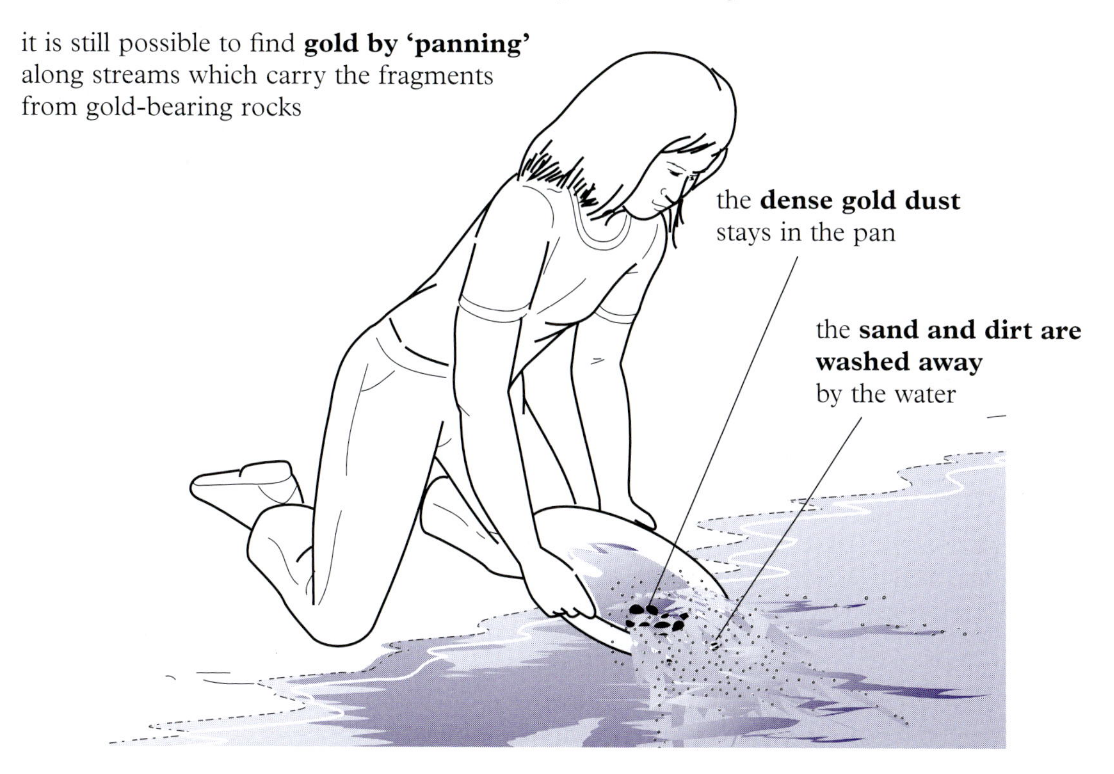
it is still possible to find **gold by 'panning'** along streams which carry the fragments from gold-bearing rocks
the **dense gold dust** stays in the pan
the **sand and dirt are washed away** by the water

Fairly reactive metals

All the metals between carbon and copper in the list fall into this group. All these metals exist as **compounds** in the rock. First the compound has to be removed from the rock. Then the metal is freed from the compound by mixing the compound with carbon and then heating the mixture. This method only works for metals **below carbon** in the reactivity series.

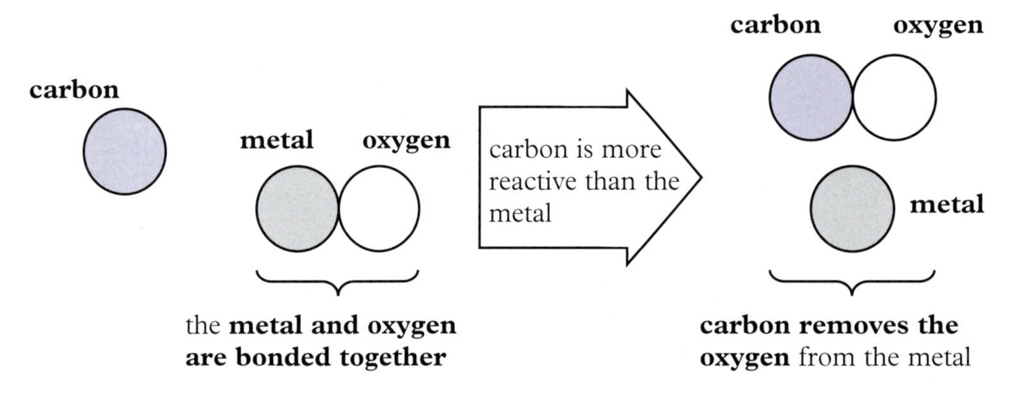
carbon
metal
oxygen
carbon is more reactive than the metal
carbon
oxygen
metal
the metal and oxygen are bonded together
carbon removes the oxygen from the metal

In this way the metal is freed. We say the compound has been **reduced** to the metal. Carbon is the **reducing agent**.

Extracting iron

Iron is obtained from iron ore which is known as haematite. Haematite contains the metal compound iron oxide. The oxygen is taken away from iron oxide in a blast furnace, leaving behind iron. Carbon monoxide is formed inside the furnace. The carbon monoxide is used to reduce the iron oxide to iron. Limestone is added to help remove impurities.

A blast furnace is a 30 metre high tower lined with brick. Blasts of hot air are blown into the bottom of the furnace.

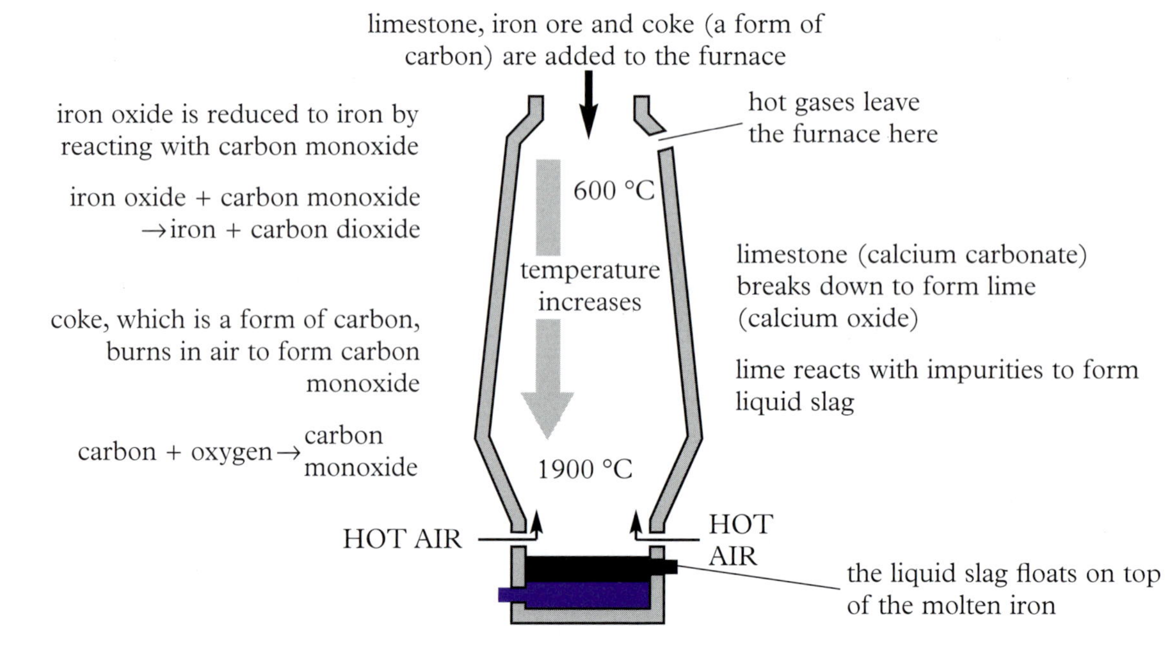
limestone, iron ore and coke (a form of carbon) are added to the furnace
iron oxide is reduced to iron by reacting with carbon monoxide
iron oxide + carbon monoxide →iron + carbon dioxide
coke, which is a form of carbon, burns in air to form carbon monoxide
carbon + oxygen→ carbon monoxide
HOT AIR
hot gases leave the furnace here
600 °C
temperature increases
1900 °C
limestone (calcium carbonate) breaks down to form lime (calcium oxide)
lime reacts with impurities to form liquid slag
HOT AIR
the liquid slag floats on top of the molten iron
the molten iron drips down the furnace and collects at the bottom

The iron from the furnace is called pig iron or **cast iron**. This contains about 4% carbon, which makes it brittle. This means that if you dropped cast iron it might break. Cast iron is a strong, hard material which means it is not easily bent or dented. Pure iron is softer and more bendy.

Uses of iron

Iron is a cheap and useful metal. Iron's biggest problem is that it rusts.

Iron is made into steels. These steels are made from iron with small amounts of other elements mixed in. Most steels are made by blowing oxygen through the molten iron to burn off some of the carbon from cast iron and adding other metals. Different steels have different properties.

Mixtures of a metal with another element are called **alloys**. The table below shows some alloys containing iron.

Name of steel	What it is made from	Special properties
Mild steel	Iron mixed with a trace of carbon (less than 0.5% carbon in the alloy)	Can be bent
Hard steel	Iron mixed with carbon (about 0.5 – 1.0% carbon in the alloy)	Hard and rigid (does not bend)
Stainless steel	74% Iron, 18% chromium, 8% nickel	Does not rust
Cast iron	Iron and carbon (2.5–4.5% carbon in the alloy)	Cheap, easily poured into a mould and cast into shapes, but it is brittle

Questions

1 Which of the following are metal compounds?
   a) copper
   b) copper sulphate
   c) hydrogen
   d) zinc chloride.

2 Answer the following and in each case explain your answer.
   a) Is a **reactive** metal more likely or less likely than an unreactive metal to be combined with another element in a chemical compound?
   b) Apart from gold, which metal is likely to be obtained as the uncombined metal?
   c) Name two metals that **could** be obtained by heating their compounds with carbon.
   d) Name a metal that **could not** be obtained by heating its compound with carbon.

3 a) Why is the furnace used for producing iron called a **blast** furnace?
   b) What is added to the furnace to provide carbon?
   c) What happens to the carbon inside the furnace?
   d) Iron oxide is reduced to iron.
      i) What does the word 'reduced' mean in this case?
      ii) What is used as the reducing agent?
   e) Why is limestone added to the furnace in the blast furnace?
   f) What does **molten** mean?

Questions (continued)

4 Steels are made from cast iron by burning off some of the carbon.
   a) Why is iron made into steels?
   b) Why is mild steel a suitable choice for car bodies and washing machines?
   c) What is the main problem when using mild steel for car bodies and washing machines?
   d) Which steel is used for knives and forks in the home? Give reasons for your answer.
   e) Which steels could **not** be used to make hammers and drill bits? Explain why these steels could not be used in this way.

Very reactive metals

Metals which are more reactive than carbon are usually obtained by melting the mineral and applying an electric current. This process splits the metal element away from the rest of the compound. The process is called **electrolysis**. Aluminium is obtained by electrolysis.

Extracting aluminium

Aluminium ore, which is known as bauxite, contains **aluminium oxide**. Pure aluminium oxide has to be separated from it. Aluminium is more reactive than carbon. This means that carbon cannot be used to take away the oxygen from the aluminium.

An electric current is used to separate the aluminium and the oxygen in an **electrolysis cell**.

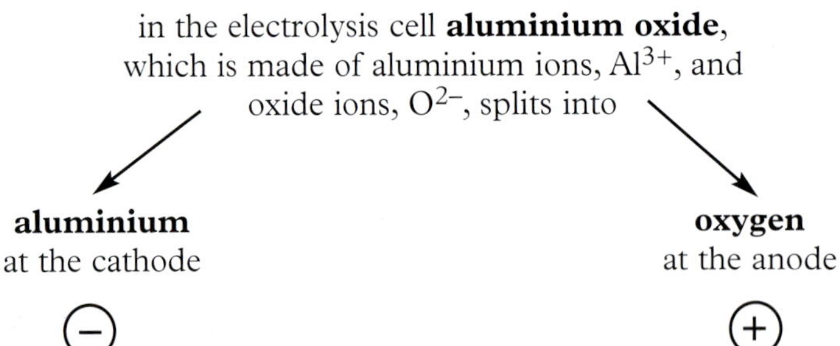
in the electrolysis cell **aluminium oxide**, which is made of aluminium ions, Al^{3+}, and oxide ions, O^{2-}, splits into
aluminium
at the cathode
−
oxygen
at the anode
+

Problems with electrolysis	Answer to the problem
Electrolysis only works if the current passes through a *liquid*. Pure aluminium oxide **melts at 2015 °C which is a very high temperature**.	Aluminium oxide is dissolved in another aluminium compound called cryolite to **form a mixture which melts at 950 °C**. This liquid can be electrolysed to produce aluminium.
Electrolysis **uses a lot of electricity**.	Aluminium plants are often situated in areas **where cheap hydroelectricity is available**.
Both electrodes are made of carbon. When the aluminium oxide is split into aluminium and oxygen, the **oxygen reacts with the hot carbon anode which burns away to carbon dioxide**.	The **anodes are replaced at regular intervals**.

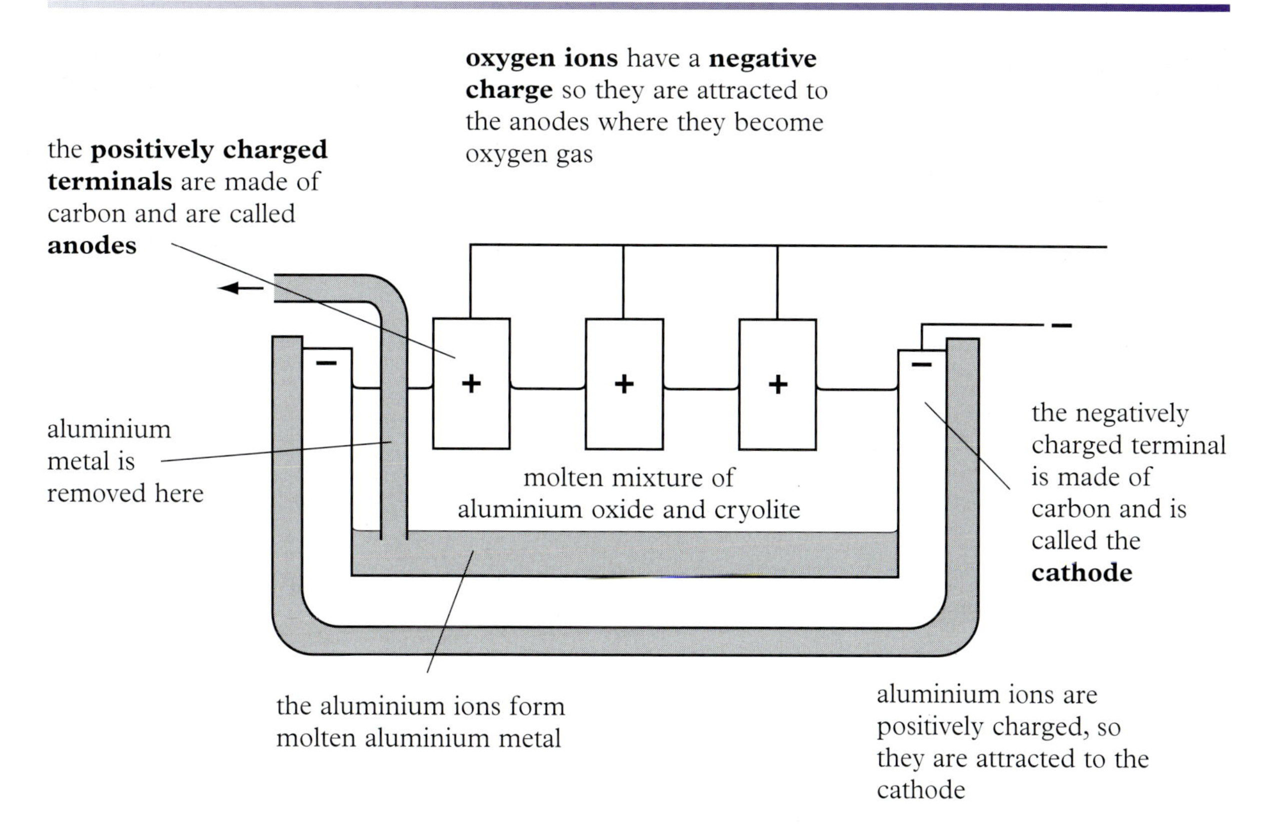

oxygen ions have a negative charge so they are attracted to the anodes where they become oxygen gas
the positively charged terminals are made of carbon and are called anodes
+
+
+
–
–
–
aluminium metal is removed here
molten mixture of aluminium oxide and cryolite
the negatively charged terminal is made of carbon and is called the cathode
the aluminium ions form molten aluminium metal
aluminium ions are positively charged, so they are attracted to the cathode

Aluminium is a lightweight (low density) metal. Aluminium conducts heat and electricity well and it can be rolled into sheets. The surface of the metal is always covered in a layer of aluminium oxide. This stops more oxygen from reacting with it and corroding the metal.

Uses of aluminium

Aluminium is used to make cooking pans, overhead electricity cables, drinks cans, window frames, aluminum foil and many other things. When aluminium is alloyed with other metals it is used to make aeroplane bodies.

Questions

1 a) In the **electrolysis cell** for producing aluminium, the mixture melted. Why is the mixture melted?
b) What is the melting temperature of pure aluminium oxide?
c) Why is it important to use a mixture, rather than pure aluminium oxide?
d) Why do the carbon anodes have to be replaced at intervals?
e) Why are aluminium ions attracted to the **negative** electrode or cathode?

2 What properties of aluminium make it suitable for making
a) a racing bike
b) a saucepan?

3 Which compound forms a layer on the surface of aluminium?

4 How does the aluminium used to make aeroplanes differ from the aluminium used to make drinks cans?

Refining copper

Another use for electrolysis is to make pure copper from copper which has other metal impurities in it. To do this, the anode is made from impure copper. Copper dissolves from the anode and the same amount of copper is deposited onto the cathode.

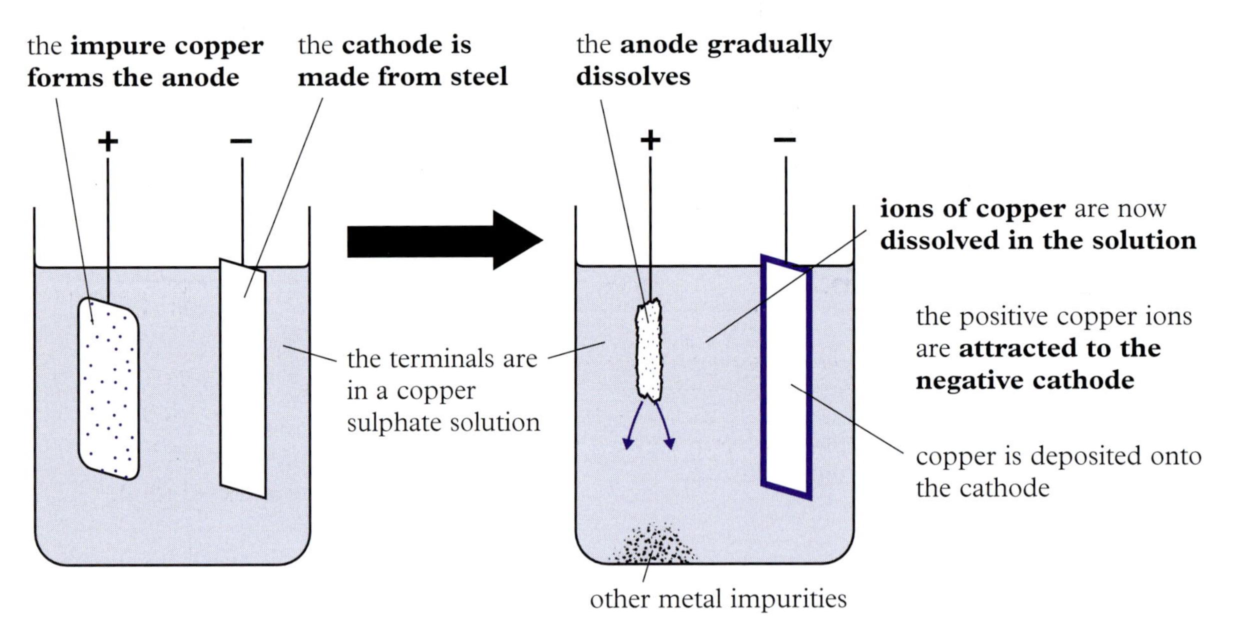
the impure copper forms the anode
the cathode is made from steel
the anode gradually dissolves
+
−
+
−
ions of copper are now dissolved in the solution
the positive copper ions are attracted to the negative cathode
the terminals are in a copper sulphate solution
copper is deposited onto the cathode
other metal impurities

Uses of copper

Pure copper is used for electricity cables because it is a very good conductor of electricity. Pure copper is used for plumbing because it is very resistant to corrosion.

Questions

1 When copper is refined by electrolysis, what happens to the mass of:
a) the anode
b) the cathode?

2 Why is copper suitable for:
a) plumbing pipes
b) electricity cables?

Electrolysis of salt

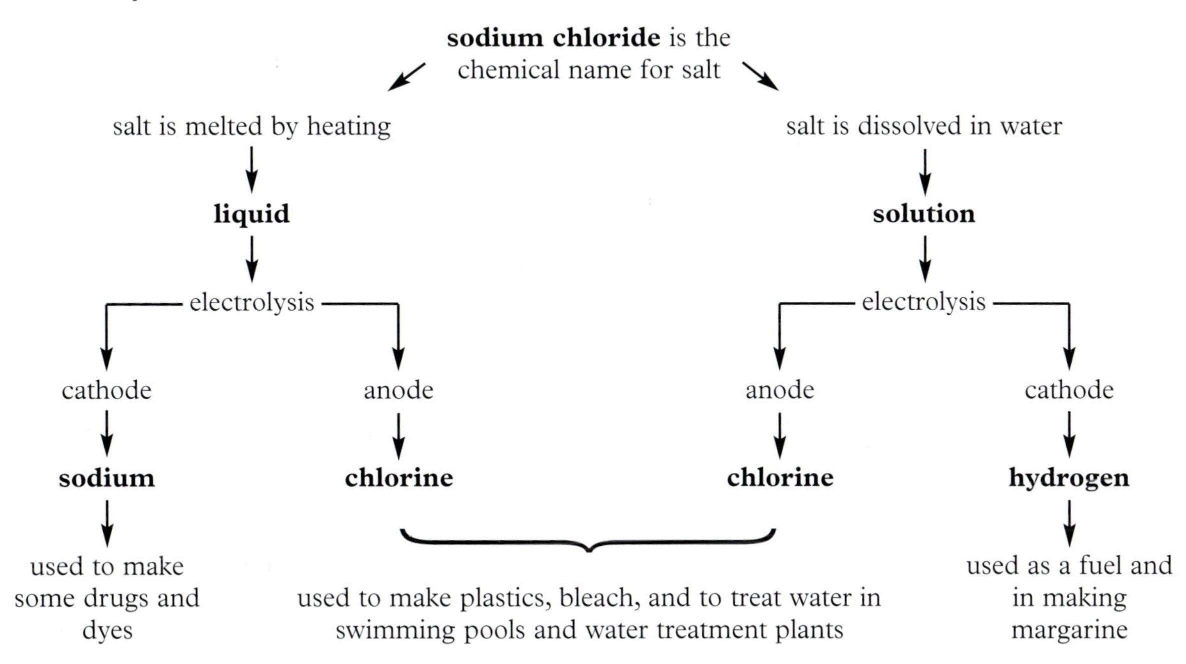
sodium chloride is the chemical name for salt
salt is melted by heating
liquid
electrolysis
cathode
sodium
used to make some drugs and dyes
anode
chlorine
salt is dissolved in water
solution
electrolysis
anode
chlorine
cathode
hydrogen
used as a fuel and in making margarine
used to make plastics, bleach, and to treat water in swimming pools and water treatment plants

Summary

Rocks are made up of chemical compounds called minerals. A metal ore is a rock that is worth mining to obtain the metal from it.

Metals can be arranged in order of their reactivity. Unreactive metals, like gold, are found uncombined with other elements. Moderately reactive metals can be obtained from their compounds by heating with carbon, which acts as a reducing agent.

Iron is produced in the blast furnace. The raw materials are iron ore, coke (a form of carbon) and limestone. Steels are mixtures of iron and other elements. These mixtures are known as alloys.

Electrolysis is used to obtain reactive metals from their compounds. Aluminium is obtained by the electrolysis of aluminium oxide dissolved in cryolite. Electrolysis is used to refine metals such as copper. Salt, which is the chemical sodium chloride, is split into its elements by electrolysis.

Key words

electrolysis The splitting up of a liquid into elements by passing an electric current through it. The metal always forms at the negative electrode.

raw materials The starting materials used to produce another chemical.

reducing agent A chemical that will remove oxygen from a compound.

refining Making pure.

End of Chapter 14 questions

1 Copy the passage and fill in the blanks from the list of words that follow. You may use each word once, more than once or not at all.

carbon   iron oxide   reduced   limestone   slag   oxygen

Iron ore contains mainly i__________ . In the blast furnace this is mixed with l__________ and coke. Coke is an impure form of c__________ . In the furnace, the iron ore is r__________ to iron. Limestone is added to get rid of impurities. The impurities and the limestone form a liquid called s__________ . The iron from the furnace contains a few percent of c__________ , which makes it brittle. In steel making, this impurity is removed by blowing o__________ through the iron.

2 Iron is extracted from its ore by reacting it with coke (a form of carbon) in the blast furnace while aluminium is extracted by electrolysis.

a) (i) Which metal, iron or aluminium, is the more reactive?
(ii) Use your answer to (i) to explain the reason for the different methods of extraction.

b) What is meant by the term electrolysis?

c) (i) Complete the word equation for the reaction which happens in the blast furnace.

iron oxide + carbon monoxide → __________ + __________

(ii) Name the reducing agent in the above equation.

d) Both aluminium and iron are used for everyday purposes.
(i) Give a reason why saucepans are made from aluminium rather than iron.
(ii) Give a reason why the jib of a crane is made from steel (a form of iron) rather than aluminium.

3 Here is part of the reactivity list of metals. Carbon has been included, although it is not a metal.

potassium	VERY REACTIVE
sodium	
calcium	
magnesium	
aluminium	
CARBON	FAIRLY REACTIVE
zinc	
iron	
tin	
lead	
copper	
silver	
gold	UNREACTIVE

a) Which of the elements in the list is most likely to be found uncombined in nature?

b) i) Give two metals that could be extracted from their oxides by heating the oxides with carbon.
   ii) Name a gas that would be produced in this process.
   iii) Write a word equation for the reaction of one of the metal oxides with carbon.

c) i) Give two metals which could **not** be extracted from their oxides by heating them with carbon.
   ii) Suggest a method that could be used to extract these metals from their ores.

4 Iron is extracted from its ore (which contains iron oxide) in the blast furnace. An outline of this is shown.

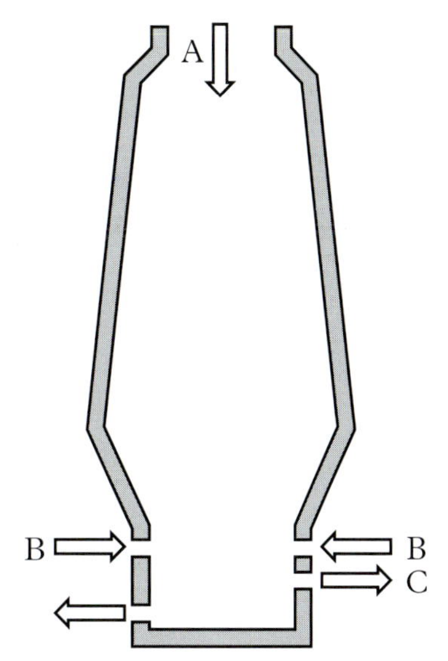
A
B
B
C

a) What three materials are added at A?
b) What enters the furnace at B?
c) What leaves the furnace at C?
d) What is slag?
e) The main reaction in the furnace is

iron oxide + carbon monoxide →
iron + carbon dioxide

   i) In the reaction carbon monoxide (CO) is oxidised to carbon dioxide (CO_2). Explain what oxidised means.
   ii) Is the iron oxide oxidised, reduced or unchanged? Explain your answer.

5 a) Aluminium is extracted from its ore, bauxite (which contains aluminium oxide), by passing electricity through the molten oxide. Iron is extracted by heating its ore with carbon. Explain the following:
   i) What is meant by the word ore?
   ii) Why must the aluminium oxide be melted before it is electrolysed?
   iii) Why can aluminium not be extracted by heating with carbon?

b) During the electrolysis, the aluminium is formed at the negative electrodes.
   i) What does this tell you about the charge on an aluminium ion?
   ii) What is the name for a negative electrode?
   iii) Which element forms at the positive electrodes?
   iv) This electrode is made of carbon. What happens to the electrode in the process?

15 Useful products from crude oil

Why crude oil is so important

Many of the things we use every day come from crude oil. Some **fuels** are made from crude oil.

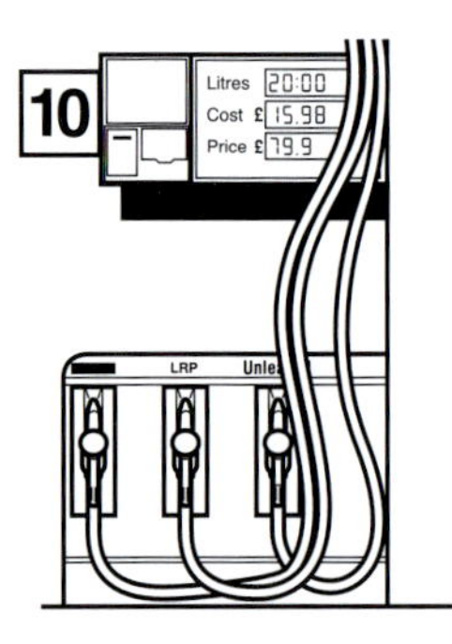
10
Litres 20.00
Cost £ 15.98
Price £ 79.9
LRP

aeroplanes, cars and lorries use **fuel** made from crude oil

GAS

we cook with **natural gas** at home, and with **butane** or **propane** when camping

Plastics are made from crude oil

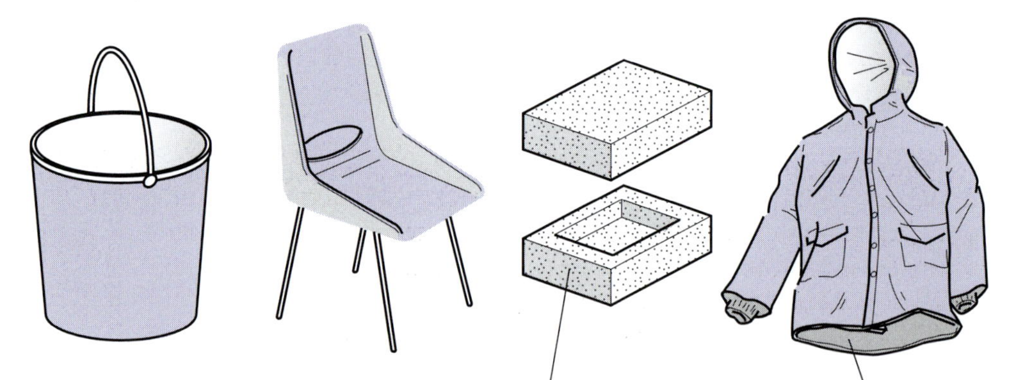

polythene and **propylene** are used for buckets, bowls and stacking chairs

polystyrene is used for packaging

PVC is used to make waterproof material

Many **paints**, **dyes**, **fabrics**, **cosmetics** and **medicines** are based on products from crude oil

P&ZUX
LOSS PAINT

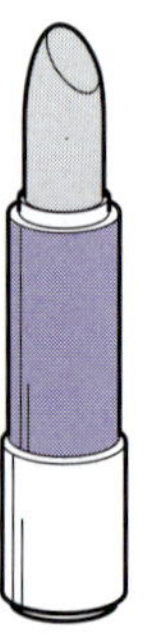

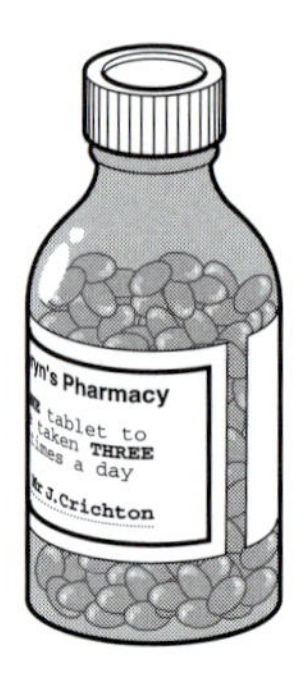
Pharmacy
tablet to
taken THREE
times a day
J.Crichton

Fuels

Fuels are **sources of energy**. We need fuels for heating and lighting our homes, for cooking, and for travelling in cars, planes and trains. Most of our fuels come from the Earth itself.

fuels that formed millions of years ago and cannot be replaced are called **non-renewable**

if a fuel can be replaced easily we say it is **renewable**

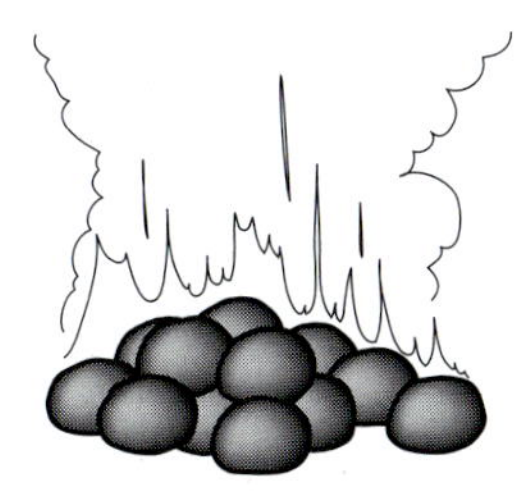

coal was formed millions of years ago from decaying plants

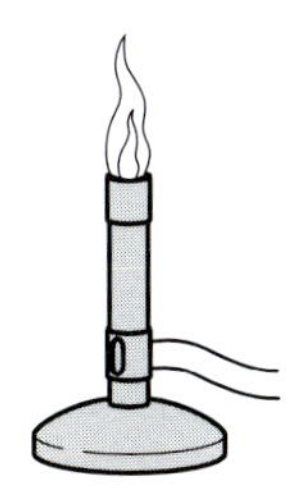

petrol and **gas** were formed millions of years ago from decaying plants and animals that lived in the sea

the **wood** from fast growing trees can be replaced quite quickly

Crude oil is a fossil fuel

Many of our fuels, like petrol and gas, come from underground deposits of crude oil. We call crude oil a **fossil fuel** because it was formed from plants and animals which lived and died millions of years ago. Fossil fuels are **non-renewable** because once we use them up, we cannot replace them.

Questions

1 a) Name four things found in the home that are made from crude oil.
b) How did fossil fuels get their name?
c) Why are they non-renewable?
d) Explain why fast-growing trees are a renewable fuel.

How crude oil formed

Crude oil was made mostly from microscopic plants and animals found in the sea. These plants and animals are called **plankton**.

millions of years ago **plankton died** and fell to the seabed

sand and mud fell in layers burying the remains of the plankton

pressure from all the layers and the **high temperatures** found deep below the Earth's surface changed the plankton into oil and gas

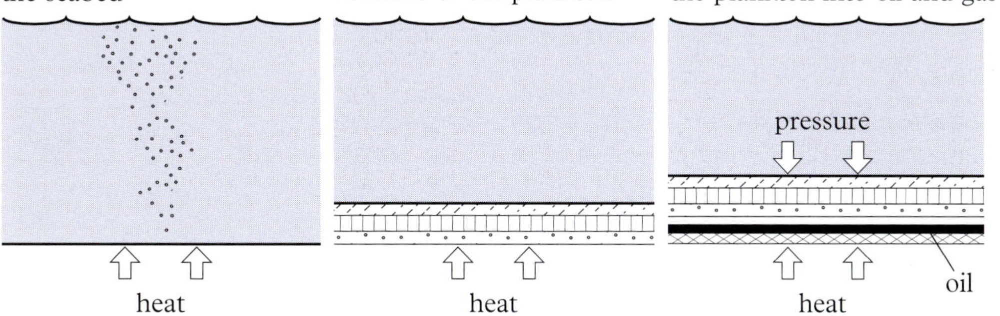
heat
heat
pressure
heat
oil

Most of the gas and oil floated upwards towards the water. The oil rose to the surface, through tiny holes called pores in the rock. Some gas and oil became trapped underground when layers of rock folded or tilted. We can drill down to reach this oil and gas and get it out.

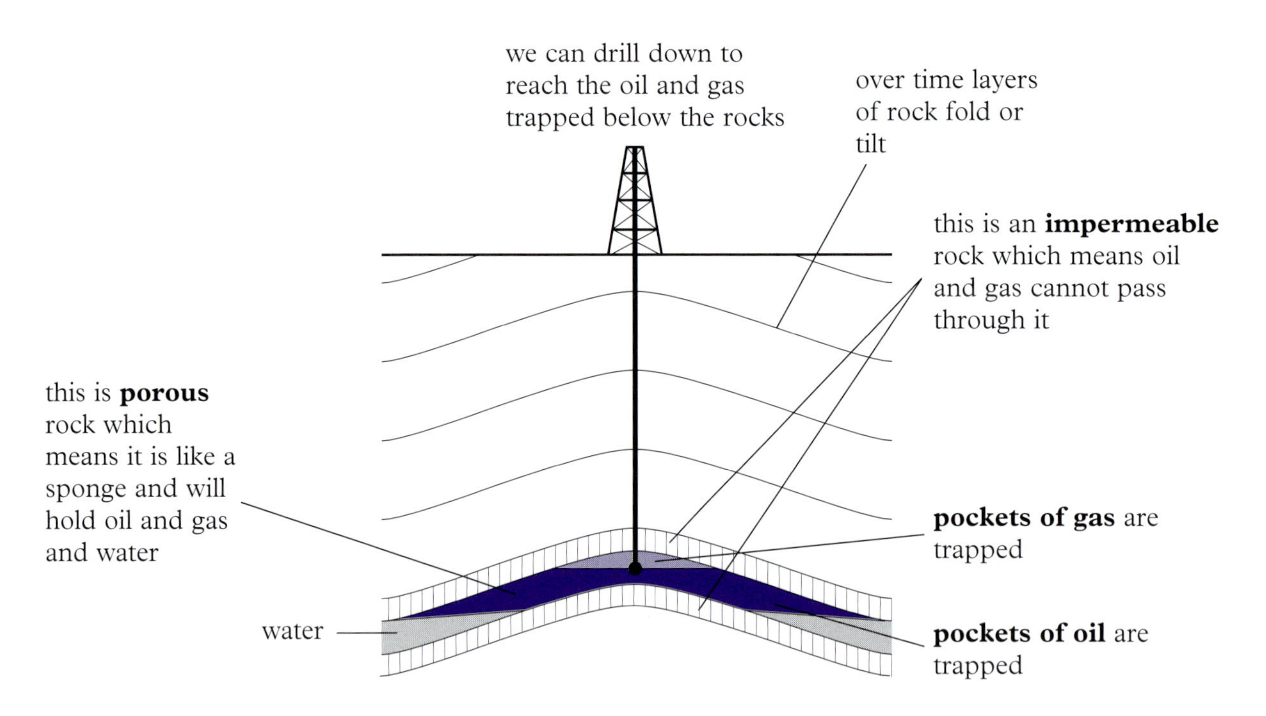
we can drill down to reach the oil and gas trapped below the rocks
over time layers of rock fold or tilt
this is an **impermeable** rock which means oil and gas cannot pass through it
this is **porous** rock which means it is like a sponge and will hold oil and gas and water
pockets of gas are trapped
water
pockets of oil are trapped

With movements of the Earth, some of these rocks were lifted above sea-level.

We have now used up much of the oil and gas which were trapped under land. This is why we now explore the sea bed to find gas and oil.

Questions

1 Explain how plankton were changed into oil under the sea.

2 How does some of the oil and gas get trapped?

3 You may have heard of a gusher, when oil spurts out of an oil well of its own accord. Explain how this might happen. Look at the diagram above and use what you know about gases.

The chemistry of crude oil

Crude oil is a dark brown liquid. It is a **mixture** of many liquids, and also has gases and solids dissolved in it.

All life is based on the elements carbon and hydrogen. After millions of years, the tiny plankton are turned into substances made up of carbon and hydrogen. Substances containing carbon and hydrogen only are called **hydrocarbons**.

The family of hydrocarbons

Crude oil is a mixture of hydrocarbons. The hydrocarbons can contain from one carbon atom up to several hundreds.

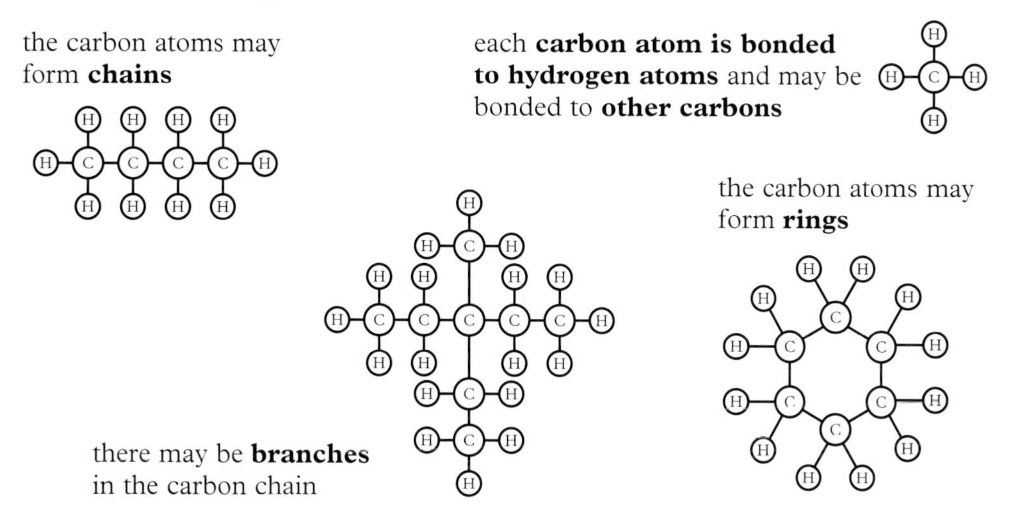
the carbon atoms may form **chains**
each **carbon atom is bonded to hydrogen atoms** and may be bonded to **other carbons**
the carbon atoms may form **rings**
there may be **branches** in the carbon chain

Separating the hydrocarbons in crude oil

Crude oil is not very useful as a mixture. So, the mixture of hydrocarbons is **separated into groups or fractions** of hydrocarbons whose molecules are of similar sizes. Hydrocarbons are separated **by boiling**, because different hydrocarbons have different boiling temperatures. The greater the number of carbon atoms in the molecule, the higher the boiling temperature of the hydrocarbon.

Crude oil is a mixture and the separation of a mixture is a **physical change** because no chemical bonds are broken and no new substances are made.

Separating the fractions of crude oil in the laboratory

The fractions of crude oil are separated by distillation. In the laboratory the distillation of oil is done in the fume cupboard because the **fumes could be harmful**. The process shown below is called **fractional distillation**.

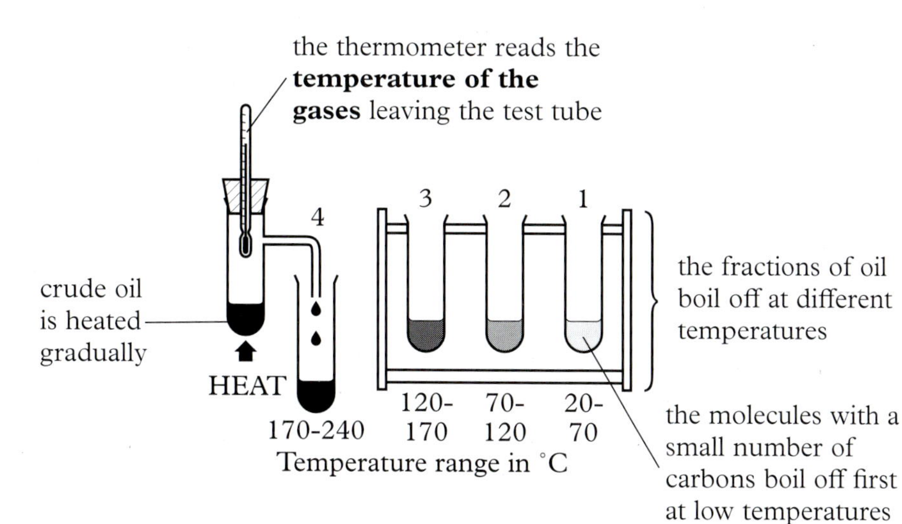
the thermometer reads the **temperature of the gases** leaving the test tube
crude oil is heated gradually
HEAT
4
3
2
1
170-240
120-170
70-120
20-70
Temperature range in °C
the fractions of oil boil off at different temperatures
the molecules with a small number of carbons boil off first at low temperatures

In industry fractional distillation is done on a huge scale. The next diagram shows a fractional distillation column being used in the oil industry.

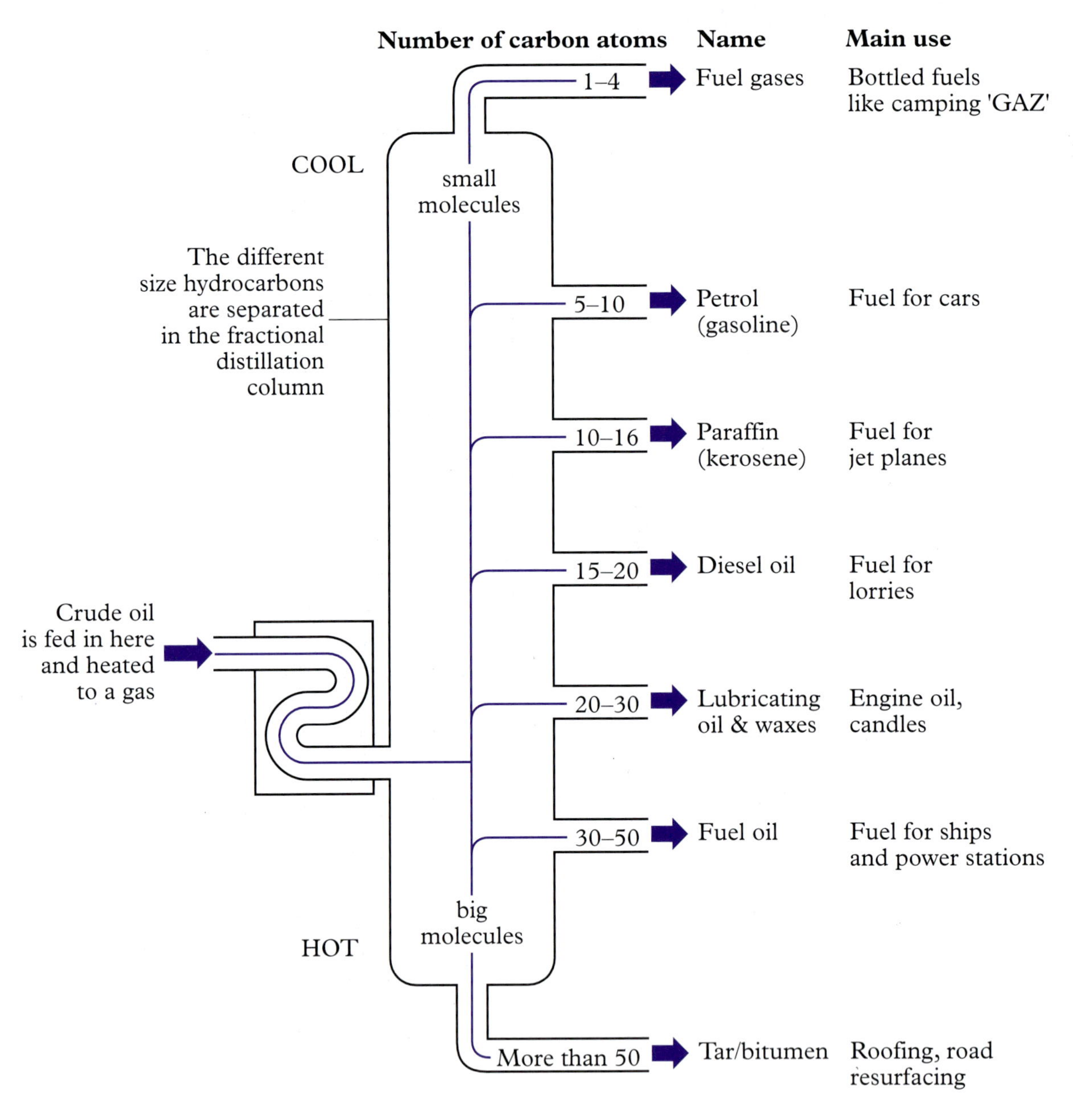
Number of carbon atoms
Name
Main use
1–4
Fuel gases
Bottled fuels like camping 'GAZ'
COOL
small molecules
The different size hydrocarbons are separated in the fractional distillation column
5–10
Petrol (gasoline)
Fuel for cars
10–16
Paraffin (kerosene)
Fuel for jet planes
15–20
Diesel oil
Fuel for lorries
Crude oil is fed in here and heated to a gas
20–30
Lubricating oil & waxes
Engine oil, candles
30–50
Fuel oil
Fuel for ships and power stations
HOT
big molecules
More than 50
Tar/bitumen
Roofing, road resurfacing

Questions

1. What is a hydrocarbon?
2. Why will heating separate the mixture of hydrocarbons?
3. Why is the process called **fractional** distillation?
4. The separation of a mixture is a **physical change**. What sort of change produces new substances?
5. Do gases have a small or large number of carbon atoms in each molecule?
6. The bitumen is left behind. Has it a high or low boiling temperature?

The different fractions

The different fractions of hydrocarbons have a pattern in their properties, according to the number of carbon atoms present in the molecule.

Hydrocarbons with fewer than four carbon atoms in the molecule are gases. Those with between five and 18 carbon atoms are liquids. The rest are solids.

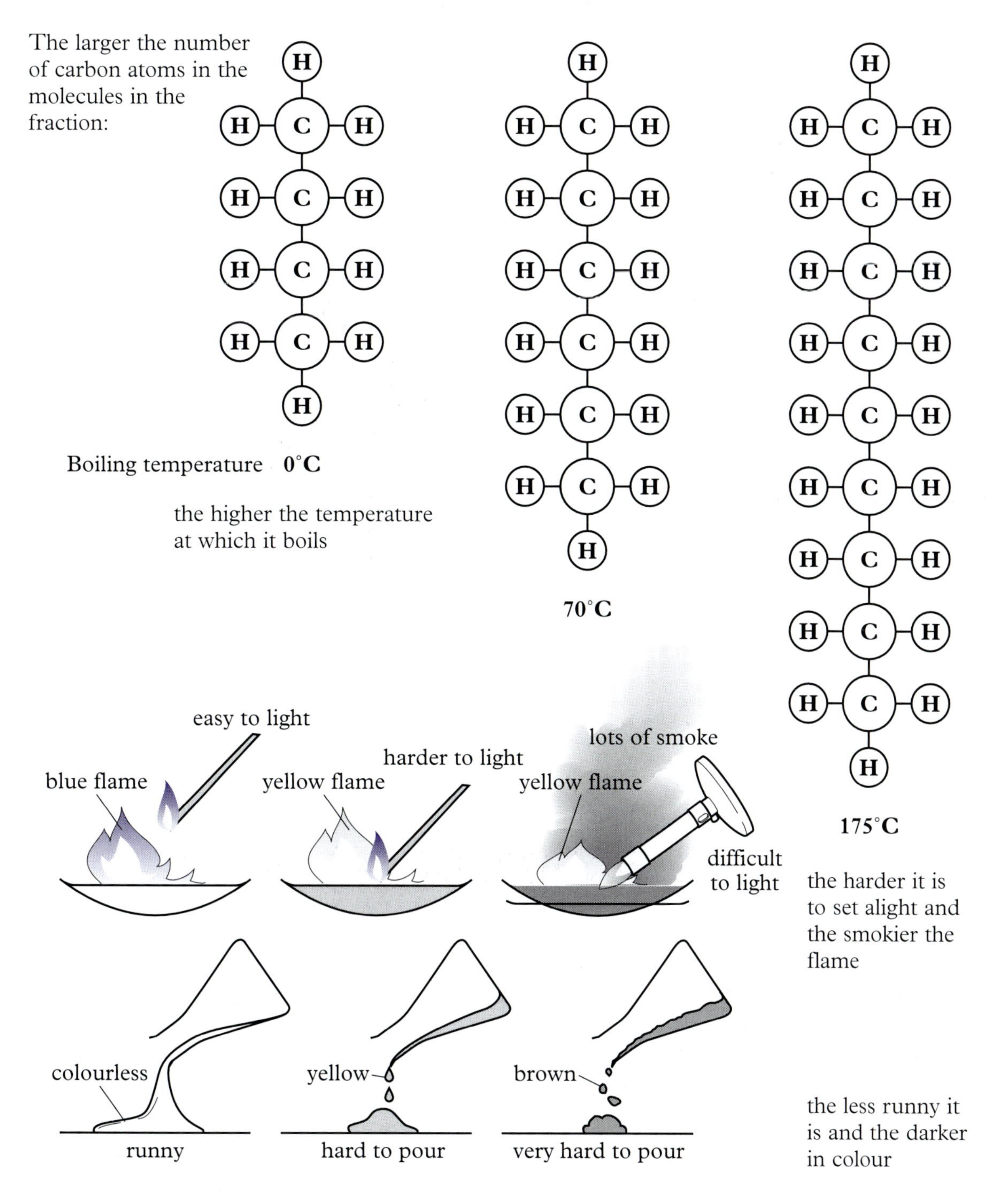
The larger the number of carbon atoms in the molecules in the fraction:
Boiling temperature 0°C
the higher the temperature at which it boils
70°C
175°C
easy to light
blue flame
harder to light
yellow flame
lots of smoke
yellow flame
difficult to light
the harder it is to set alight and the smokier the flame
colourless
yellow
brown
runny
hard to pour
very hard to pour
the less runny it is and the darker in colour

Remember that the fractions are not single substances. Each fraction contains a mixture of hydrocarbons which boils over a particular range of temperatures.

Questions

1 Look carefully at the structure of the molecules in the hydrocarbons above.

a) What do you notice about the molecules and their boiling temperatures?

b) How does the colour of each hydrocarbon change with the structure of each molecule?

c) If a fraction from the distillation of crude oil is colourless and burns easily with no smoke, you should be able to predict two other properties of the fraction. Explain what you could predict about the runniness and boiling temperature of the fraction.

Cracking

There is a huge world demand for gas as well as petrol and other liquid fuels. There is never enough of these fractions of crude oil, which contain the smaller molecules. But the fractions which contain the longer carbon chains are not in so much demand.

The oil industry gets round this problem by using heat to break down the longer chain hydrocarbons into shorter chains. The shorter chains can then be sold either as fuels or to be made into plastics. The heating is strong enough to break the chemical bonds between some of the carbon atoms.

This breaking up of the molecules is called **cracking**.

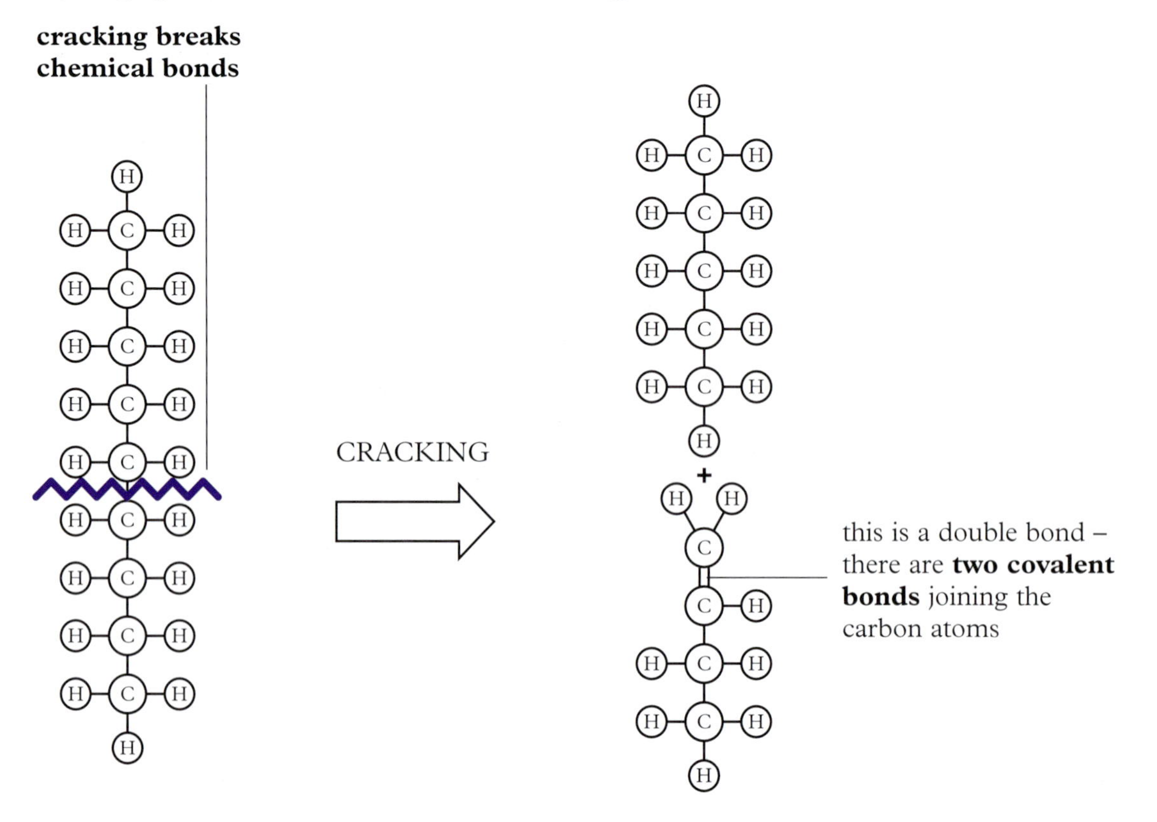
cracking breaks chemical bonds
CRACKING
this is a double bond – there are two covalent bonds joining the carbon atoms

Cracking is done using a **catalyst**. Using a catalyst causes cracking at a lower temperature. This is often called **cat cracking**. Cracking is a **chemical change** because chemical bonds are broken.

Questions

1. Which fractions of crude oil contain smaller molecules?
2. Explain how fractions which contain the longer carbon chains can be made more useful.
3. What is meant by cat cracking?
4. Why is a catalyst used in cracking?

Naming hydrocarbons

Hydrocarbons with no double bonds are called **alkanes**. Their names all end in **-ane**.

Alkanes are called **saturated** molecules because they do not have a double bond. This means that they have the maximum possible amount of hydrogen in them.

The first three alkanes are as follows:

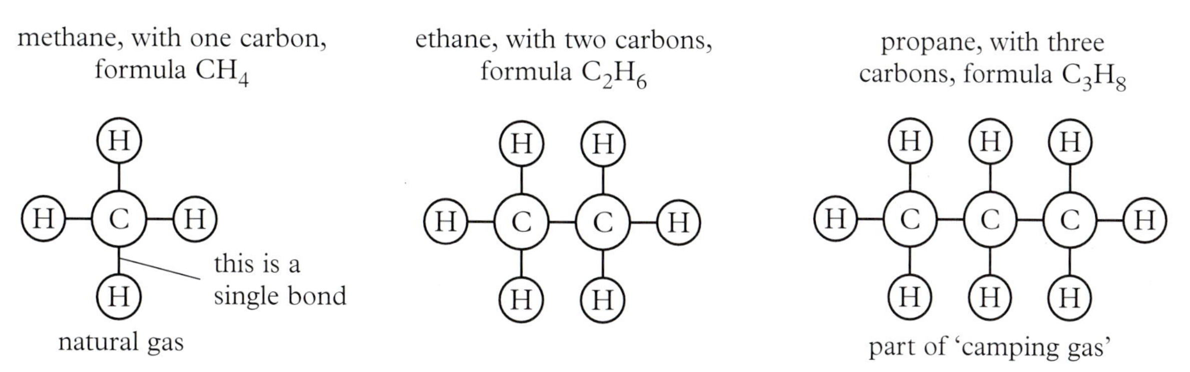
methane, with one carbon, formula CH_4
H
H
C
H
H
this is a single bond
natural gas
ethane, with two carbons, formula C_2H_6
propane, with three carbons, formula C_3H_8
part of 'camping gas'

Alkenes

Hydrocarbons which have a double bond between two carbon atoms are called **alkenes**. Their names all end in **-ene**.

The simplest alkene is **ethene**. It is an important chemical because it is used as a starting material for making many plastics.

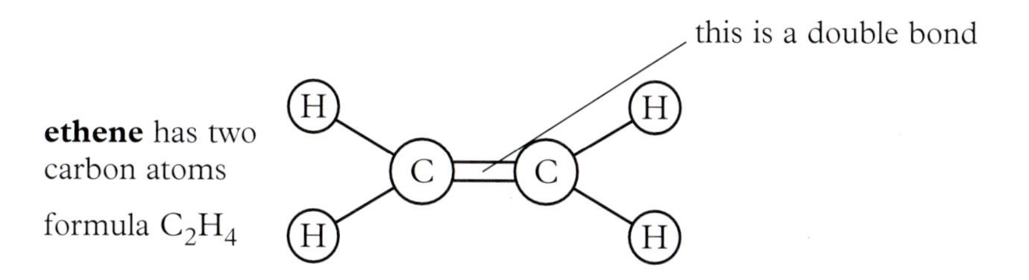
this is a double bond
ethene has two carbon atoms
formula C_2H_4
H
C
C
H

Alkenes are called **unsaturated** molecules. The double bond means alkenes have less hydrogen than an alkane with the same number of carbon atoms.

Questions

1 Why are alkanes called saturated molecules?

2 Look at the pattern of carbon and hydrogen atoms in the molecules above. The next alkane is called butane. There are four carbon atoms in a butane molecule.

a) Draw the structure of butane.
b) Write down the formula for butane.

3 Explain the difference between an alkane and an alkene.

Polymers

Since about 1930, many new materials have been developed. This means that we do not have to rely on natural materials like wool, cotton and wood any more. In fact we can make a material with just the right properties for any particular job.

Nylon, polythene, polyesters, polystyrene, pvc and hundreds more are all materials that are made from crude oil. We call them plastics because they can be moulded when they are first made.

Scientists learnt how to build large molecules from small molecules, like ethene. Ethene, which comes from crude oil, has a double bond. Ethene is more reactive than hydrocarbons with single bonds because it has a spare bond to which other atoms can add.

Polymers are formed when small molecules with a double bond are added to each other in such a way that a new material is made. The name of the material often begins with '**poly**' which means many. The small molecule that we start with is called a **monomer**.

n(monomer) → polymer

n(monomer) means 'lots of' monomers.

If we start with the monomer called ethene we can make a useful polymer.

n(ethene) → poly(ethene), called polythene

monomer (n(ethene)) — polymer (poly(ethene))

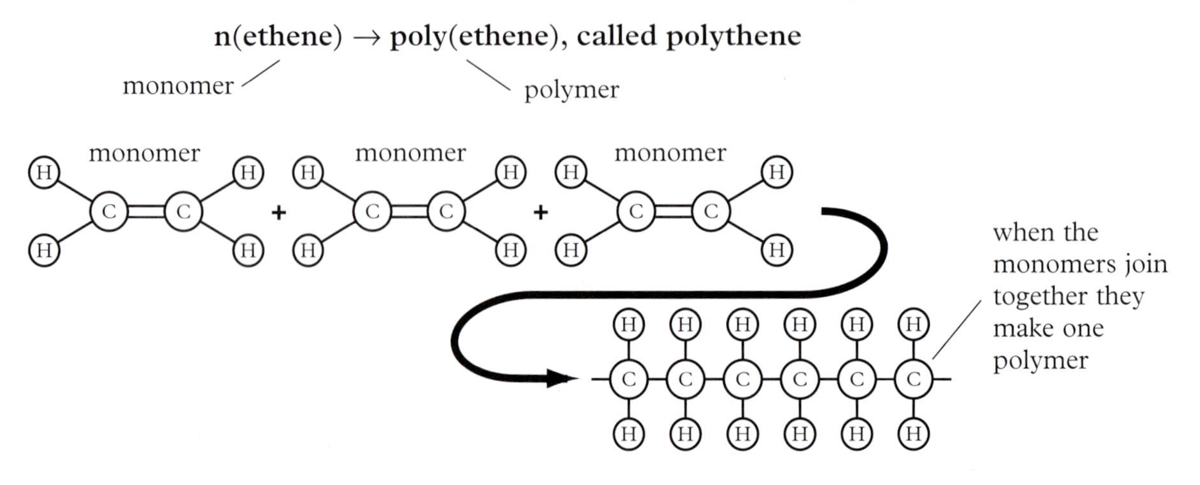

monomer
monomer
monomer
H
C
+
when the monomers join together they make one polymer

We use many different polymers every day:

Material	What the material is used for
poly(ethene)	shopping bags, washing up bowls, plastic chair seats
polystyrene	moulded packing material
polyvinylchloride which is pvc	plastic drain pipes

Oil and our environment

Oil spills

Oil is moved across the world in huge tanker ships or through underground pipes. Most of the time we do not think about its transport. But every now and then an oil spill occurs at sea and its effects can be serious:

- The sea-bird life in the area is often killed off in large numbers because the birds' feathers get covered in crude oil. This means the birds cannot fly or float.
- Fish and seals may die as they get coated in oil or are poisoned by swallowing oil.
- Seaside resorts lose money as people do not come for their holidays in areas affected by oil.
- Fisherman in the area cannot fish for their living.

Burning fossil fuels

Whenever we burn a fossil fuel in a plentiful supply of air, the carbon, C, burns to form carbon dioxide, CO_2, and the hydrogen, H, burns to form water, H_2O. As we continue to burn more and more fossil fuels, the level of carbon dioxide in the atmosphere is gradually rising. Carbon dioxide is a greenhouse gas, which means it traps the heat escaping from the Earth's surface. Our atmosphere is getting slowly warmer. This is called **global warming** and if it goes on happening a change in weather may affect where we can farm land across the planet. Sea levels will rise as polar ice caps melt and the water in the sea expands.

If we burn fossil fuels in a poor supply of air, there may not be enough oxygen to convert all the carbon in the fuel into carbon dioxide. In this case, some carbon monoxide, CO, may form. This is a poisonous gas. So it is important to ensure a good supply of air for gas heaters, and not to run car engines in a closed garage.

These are not the only problems caused by burning fossil fuels. Many fossil fuels contain a little sulphur. When this is burned, sulphur dioxide gas is formed. This combines with air and water in the atmosphere to form sulphuric acid which then falls to the Earth with rain. This is called **acid rain**. Acid rain is suspected of killing trees and fish in forests and lakes in affected areas.

Another cause of acid rain is a mixture of nitrogen oxides, often called NO_x. These are made when oxygen and nitrogen in the air combine together at high temperatures, such as inside a car engine when the fuel is burned. NO_x combines with air and water to form nitric acid.

air entering the engine contains nitrogen and oxygen
petrol in the tank contains hydrocarbons and a little sulphur
burning the petrol produces exhaust gases containing NO_x, CO_2, CO, SO_2, H_2O

Questions

1 Ethene is a monomer. We can make a polymer from ethene.

a) What is the name of the polymer we can make from ethene?

b) Give two examples of what this polymer is used for.

2 a) Why is carbon dioxide called a greenhouse gas?

b) Why is global warming a problem?

c) Acid rain can be formed from two acids. Name both acids.

Summary

Crude oil is a mixture of different hydrocarbons. Oil is a fossil fuel. It is separated into fractions containing hydrocarbons of different lengths by fractional distillation. The long chain hydrocarbons are made into shorter, more useful hydrocarbons by cracking.

Alkanes are saturated hydrocarbons with single bonds. Alkenes are unsaturated hydrocarbons because they have a double bond. Ethene, C_2H_4, is the simplest alkene. It is used to make the polymer, polythene. Polymers are materials which are made with crude oil as a starting material.

Key words

alkanes Hydrocarbons with single bonds only.

alkenes Hydrocarbons with a double bond.

cracking A process in which long chain hydrocarbons are heated to break them down into shorter, more useful hydrocarbons.

fossil fuel A fuel produced from plants and animals which lived millions of years ago.

hydrocarbon A molecule made of hydrogen and carbon only.

fractional distillation A separation process in which a mixture of liquids is heated and the fractions which boil off over different temperature ranges are collected.

polymerisation A process in which small molecules with a double bond add to each other to make a large molecule called a polymer.

End of Chapter 15 questions

1 a) In a Bunsen burner, natural gas (methane, CH_4) burns in a plentiful supply of air.
   i) Name the two elements in methane.
   ii) Name the two products of the reaction when methane burns in air.
 b) One of the products of the reaction is a gas, which causes a problem because its level in the atmosphere is increasing.
   i) Name the gas which is causing a problem.
   ii) What problem does the gas you have named cause?
   iii) Briefly describe what happens because of this problem.
 c) Methane is a fossil fuel and is found in deposits below the ground.
   i) Explain what is meant by the term fossil fuel.
   ii) Describe how fossil fuels were formed.
   iii) Name two other fossil fuels.
 d) Fossil fuels may be solids, liquids or gases. Gaseous fuels like methane are very convenient for use in the home. Explain why liquid fuels rather than gaseous fuels are usually used in cars and other vehicles.

2 The substances in the following list are all fossil fuels.

coal   natural gas   diesel   petrol

 a) i) Which two of the fossil fuels come from crude oil?
   ii) Briefly describe the process by which these two are separated from crude oil.
   iii) What is the name of the process used to separate fossil fuels from crude oil?
 b) Natural gas, diesel and petrol all contain compounds called hydrocarbons. What two elements are present in hydrocarbons?
 c) What two products are formed when hydrocarbons burn in a plentiful supply of air?
 d) When hydrocarbons burn in a poor supply of air, a poisonous gas may be formed. Name the gas which is formed.

3 Fractional distillation of crude oil was carried out in the laboratory. In one experiment, five fractions were collected as shown in the table below.

Fraction	Collecting temperature / °C	Colour	Smokiness of flame
1	up to 50	colourless	no smoke
2	50–100	A	a little smoke
3	100–150	yellow	B
4	150–200	yellow–brown	very smoky
5	200–250	brown–black	thick, black smoke

 a) What word would replace the letter A in the table?
 b) How would you describe the smokiness of the flame, to replace the letter B in the table?
 c) Give two other properties of the fractions which increase from fraction 1 to fraction 5 in the table.

d) The fractions contain hydrocarbons. Explain the term hydrocarbon.
e) What happens to the number of carbon atoms in the molecules as you read the table from fraction 1 to fraction 5?

4

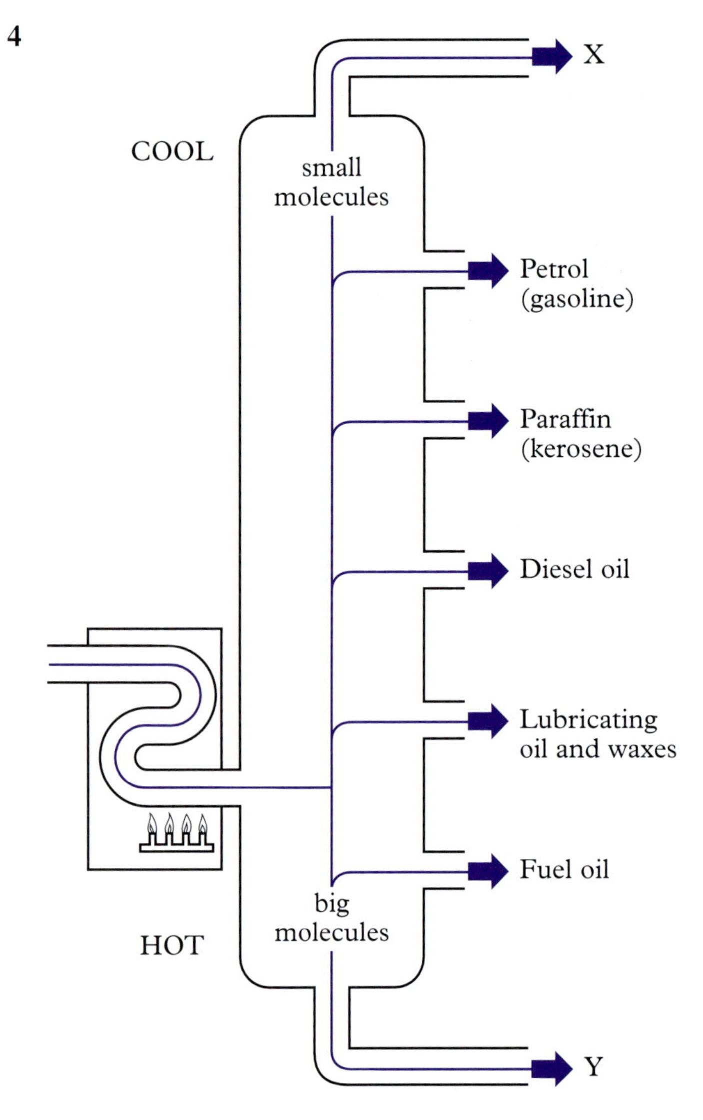
X
COOL
small
molecules
Petrol
(gasoline)
Paraffin
(kerosene)
Diesel oil
Lubricating
oil and waxes
Fuel oil
big
molecules
HOT
Y

The diagram shows how crude oil is fractionally distilled.

a) Name the fraction collected at X.
b) Give one use for the fraction labelled X.
c) Name the fraction collected at Y.
d) How could the fraction labelled Y be used?
e) Which fraction is in most demand commercially?
f) What is the main difference between the molecules in the petrol fraction and those in the lubricating oil fraction?
g) 'Cat' cracking is a process which is carried out on some of the fractions obtained from the distillation of crude oil.
   i) What does cat cracking do to long chain hydrocarbon molecules?
   ii) Why is the process called cat cracking?

5 Petrol, diesel and paraffin are described as non-renewable fuels.

a) Explain what is meant by the term non-renewable.

b) Petrol, diesel and paraffin are fuels made from crude oil which is found in underground deposits.

i) Explain where crude oil comes from.

ii) What was the original source of the energy stored in these fuels?

c) Name one other fuel that comes from crude oil.

6 The molecules below are all hydrocarbons.

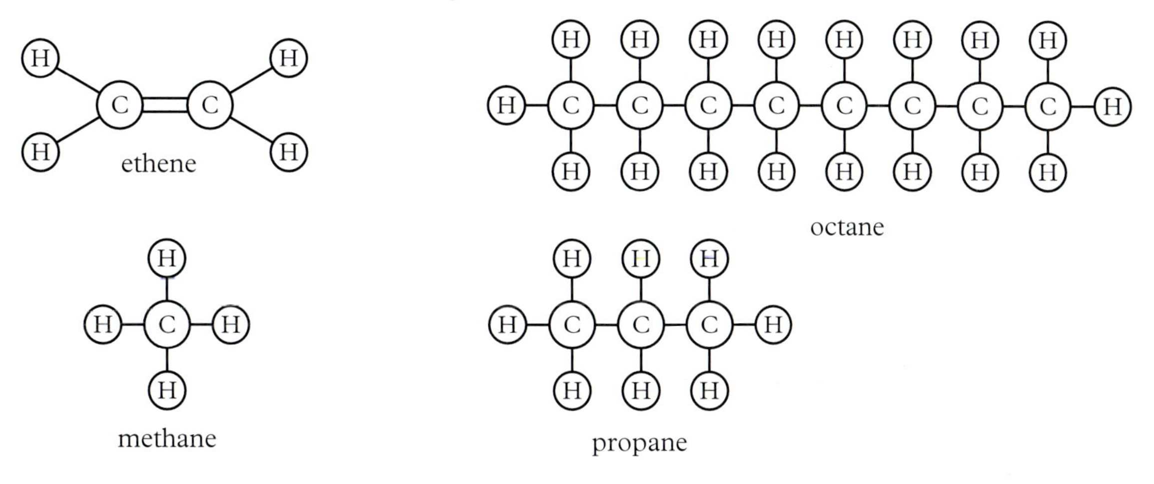
ethene
octane
methane
propane

a) Which molecule is described as unsaturated?

b) Which molecule is a liquid?

c) Which molecule is found in camping gas?

d) Which molecule is natural gas?

e) Which molecule is used to make polythene?

f) The formula of ethene is C_2H_4. Write down the formulae of propane, methane and octane.

g) Draw the formula of the hydrocarbon with two carbon atoms and no double bonds.

16 Rates of reaction

A bomb explodes because a very, very fast chemical change has taken place. Some chemical changes are very slow. A can rusts because very slow chemical changes are taking place. These different reactions take place at different **rates**.

an apple rotting away is a **slow reaction**

setting off a firework starts a **fast reaction**

In industry, chemical reactions are used to produce useful materials. Chemists need to control how quickly the reactions take place. The faster the reaction, the more quickly the materials are produced. If the reaction is too fast, it could get out of control and be dangerous.

The main things that change the rate of a reaction are:

- **Surface area** if a solid is used – the surface area is the area on the outside of the solid.
- **Concentration** if a solution is used – the concentration tells you how much solid is dissolved in a solution.
- **Temperature** – the temperature measures how hot the reaction is.
- **Catalysts** – catalysts are substances which are added to speed up a reaction but are not used up themselves.

Investigations into changing the rate of a reaction

There are many ways that you can keep track of a chemical reaction. The easiest ones to follow are the ones that give off a gas. For this sort of reaction, you can **measure how quickly the gas is given off**. The faster the gas is given off the faster the reaction is taking place.

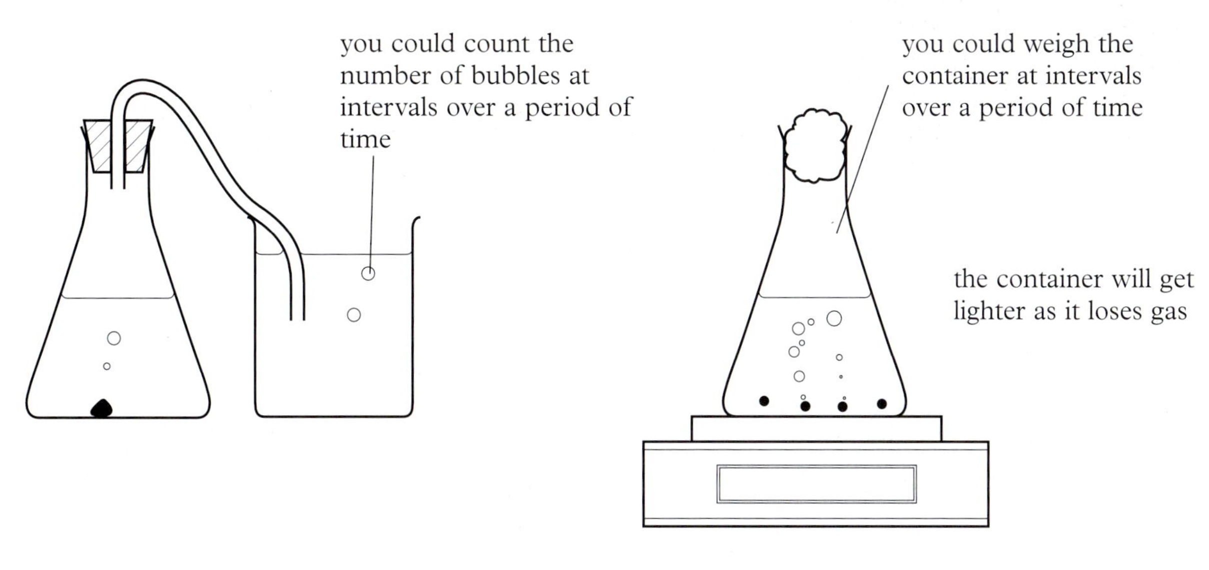
you could count the number of bubbles at intervals over a period of time
you could weigh the container at intervals over a period of time
the container will get lighter as it loses gas

Alternatively you could collect the gas and measure the increase in volume over time.

The effect of surface area

Marble chips react with hydrochloric acid, and the gas carbon dioxide is given off. Marble chips are calcium carbonate.

calcium carbonate + hydrochloric acid ⟶ calcium chloride + water + carbon dioxide

$$CaCO_3(s) + 2HCl(aq) \longrightarrow CaCl_2(aq) + H_2O(l) + CO_2(g)$$

We can use this reaction to find the effect of changing the surface area of a solid on the rate of a reaction. When this apparatus was set up the cylinder was full of water.

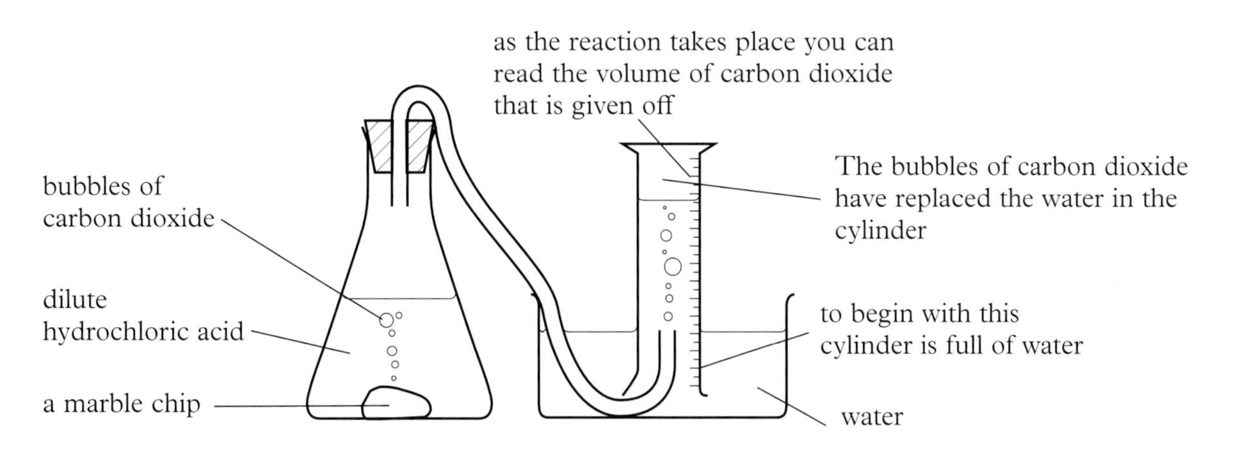
as the reaction takes place you can read the volume of carbon dioxide that is given off
bubbles of carbon dioxide
dilute hydrochloric acid
a marble chip
The bubbles of carbon dioxide have replaced the water in the cylinder
to begin with this cylinder is full of water
water

Changing the surface area

You need to do the experiment three times. You must change only one thing from one experiment to the next, so that you know for sure what is making a difference. Each time you start with a marble chip of the same size. You also always use the same volume of fresh acid each time.

In this experiment you only change the surface area of the chip.

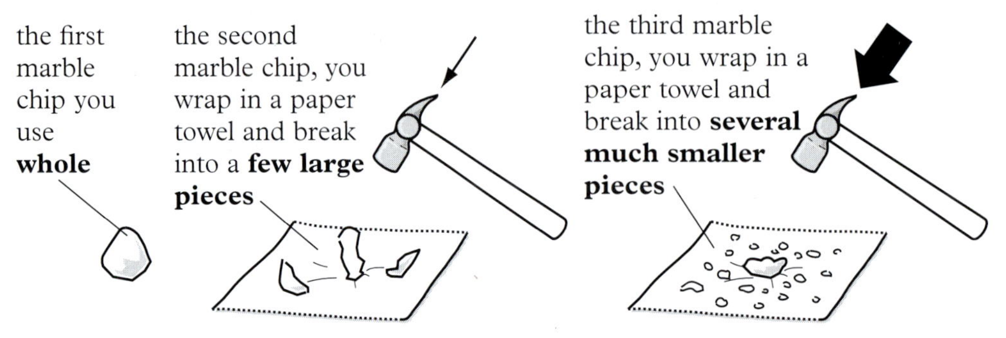
the first marble chip you use **whole**
the second marble chip, you wrap in a paper towel and break into a **few large pieces**
the third marble chip, you wrap in a paper towel and break into **several much smaller pieces**

Breaking a large marble chip into smaller pieces increases the surface area. The smaller the pieces of marble chip, the greater the surface area.

The three chips shown above were used for this experiment. The total amount of gas given off by each was noted every minute for six minutes.

The results for each piece of marble were plotted on the graph below.

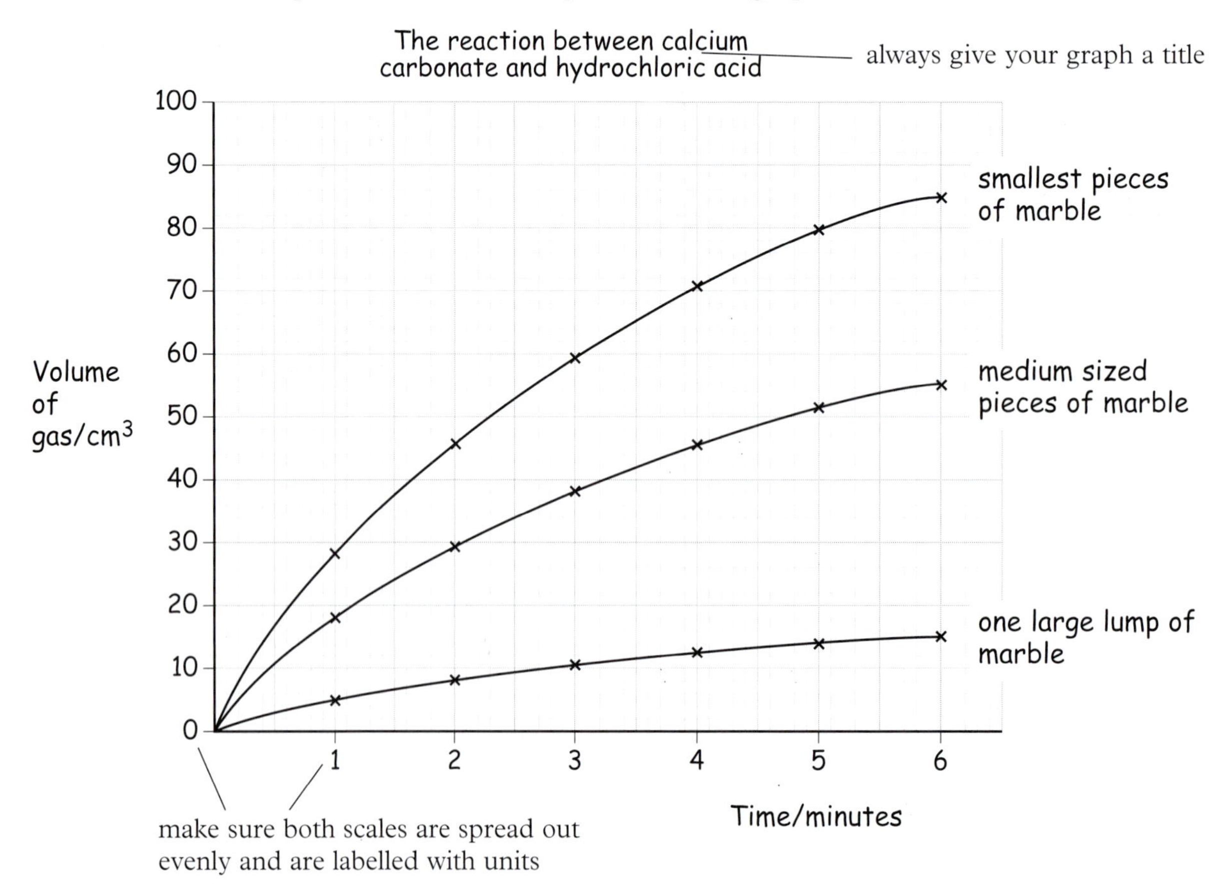
The reaction between calcium carbonate and hydrochloric acid
always give your graph a title
Volume of gas/cm3
100
90
80
70
60
50
40
30
20
10
0
1
2
3
4
5
6
Time/minutes
smallest pieces of marble
medium sized pieces of marble
one large lump of marble
make sure both scales are spread out evenly and are labelled with units

The steeper the slope of the graph, the faster the reaction took place. The graphs show that the greater the surface area, the faster the reaction.

Questions

1 In the experiment with the marble chips, why should you wrap the marble chip in a paper towel before you hit it with a hammer?

2 Each side of a cube is 1 cm so the area of each face of the cube is 1 cm^2. Eight cubes with 1 cm^2 faces were arranged to make three different shapes.

a) What is the total surface area of shape 1?

b) What is the total surface area of shape 2?

c) What is the total surface area of shape 3?

3 Look at the graph on page 168.

a) How much gas was collected after three minutes using:

i) one large marble chip

ii) medium sized pieces of marble

iii) small pieces of marble?

b) What does the result tell you about how surface area affects the rate of reaction?

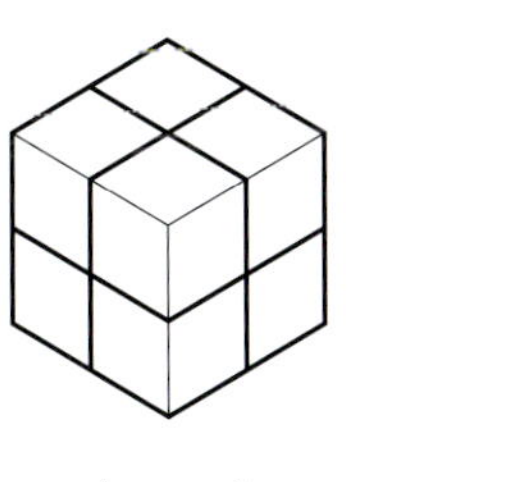

shape 1

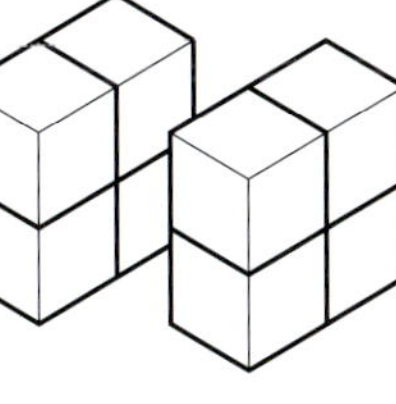

shape 2

shape 3

The effect of changing the concentration of a solution

The reaction between magnesium and hydrochloric acid gives off the gas hydrogen:

magnesium + hydrochloric acid → magnesium chloride + hydrogen

$$Mg(s) + 2HCl(aq) \rightarrow MgCl_2(aq) + H_2(g)$$

We can set up the same apparatus used to investigate the effects of changing the surface area. In this experiment we change the concentration of the acid.

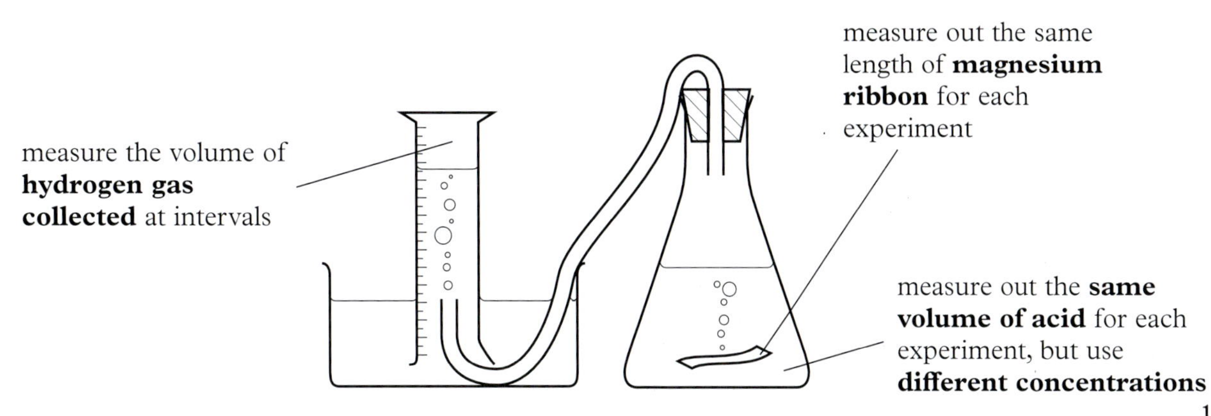
measure out the same length of **magnesium ribbon** for each experiment
measure the volume of **hydrogen gas collected** at intervals
measure out the **same volume of acid** for each experiment, but use **different concentrations**

For the first experiment use dilute hydrochloric acid (0.5 mol/litre). For the second experiment use normal concentration hydrochloric acid (1 mol/litre). For the third experiment use more concentrated hydrochloric acid (2 mol/litre).

Here are graphs of the results of an experiment investigating the effects of changing the concentration of a solution. Each experiment continued until all the magnesium had been used up.

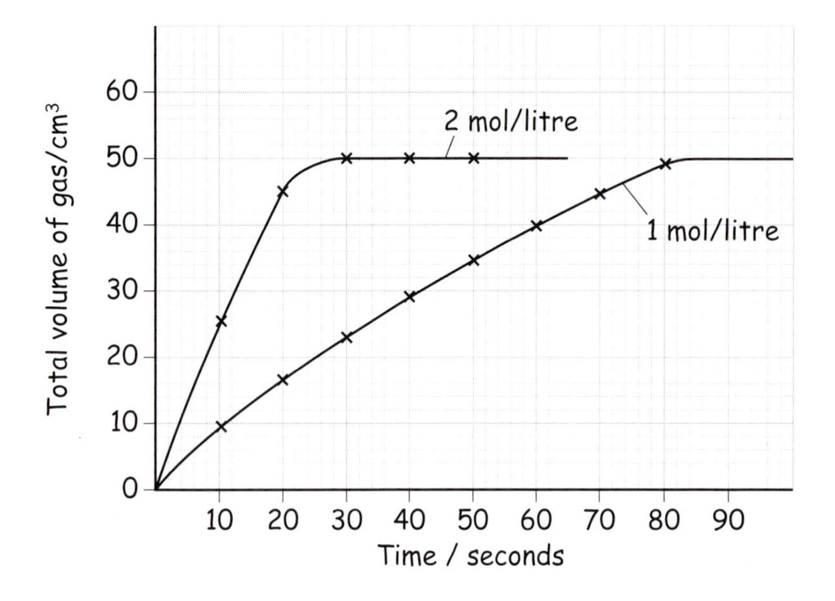
Total volume of gas/cm³
60
50
40
30
20
10
0
2 mol/litre
1 mol/litre
10 20 30 40 50 60 70 80 90
Time / seconds

The graphs above show that the greater the concentration of the solution, the faster the reaction.

Questions

1 a) Why is it important to use the same length of magnesium ribbon and the same volume of acid for every experiment in the above investigation?

b) Look at the graph above.

i) Give the graph a title.

ii) What is the final volume of gas collected?

iii) Why is the final volume the same for both experiments?

iv) On a sketch of the graphs, add a line for the dilute acid (0.5 mol/litre).

The effect of temperature

If a gas is produced in a reaction and the gas can be collected, you could use the same method as before to investigate the effect of temperature.

Some reactions do not produce a gas that can be collected easily. For example when sodium thiosulphate solution reacts with dilute hydrochloric acid, sulphur is made. In this case the sulphur is a fine powder.

sodium thiosulphate + hydrochloric acid → sulphur + sodium chloride + sulphur dioxide + water

The sulphur dioxide gas remains dissolved in the water and cannot be collected.

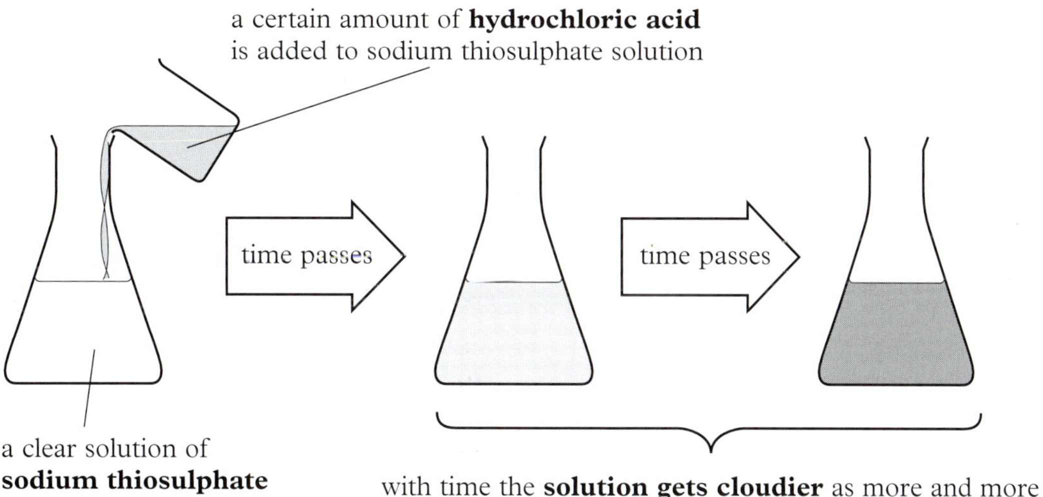

a certain amount of **hydrochloric acid** is added to sodium thiosulphate solution
time passes
time passes
a clear solution of **sodium thiosulphate**
with time the **solution gets cloudier** as more and more powdered sulphur is suspended in the solution

You can follow this reaction by timing how long it takes to get to a particular level of cloudiness. This is what happens in the 'disappearing cross' experiment.

Using the disappearing cross experiment to investigate the effect of temperature

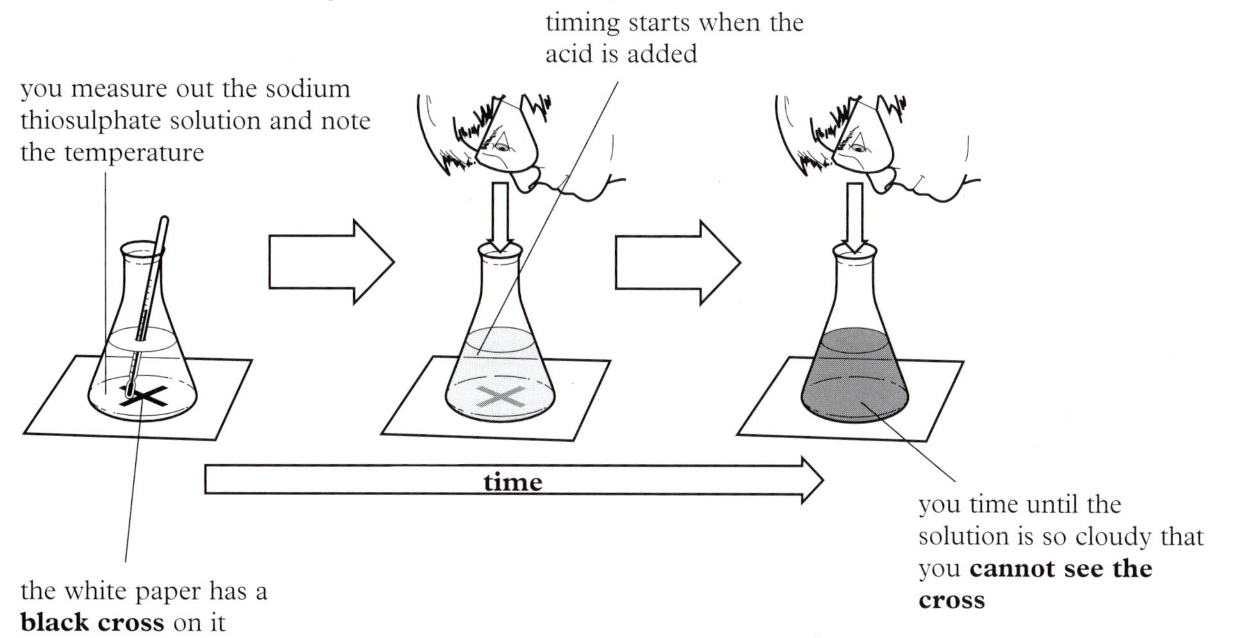

timing starts when the acid is added
you measure out the sodium thiosulphate solution and note the temperature
time
you time until the solution is so cloudy that you **cannot see the cross**
the white paper has a **black cross** on it

Then you do the whole experiment again with fresh solutions warmed up to a different temperature.

The results of an experiment like this one are shown in the table below.

Temperature of mixture / °C	Time for cross to disappear / seconds
15	100
35	25
55	7
75	2

The table shows that the higher the temperature, the faster the reaction.

The results show that if we cool things down, chemical reactions will go more slowly. This explains how fridges and freezers keep food fit to eat for longer. At lower temperatures, the reactions which make food 'go off' take place more slowly.

Questions

1 You are investigating the reaction between magnesium ribbon and hydrochloric acid. This reaction gives off hydrogen.

a) How would you find how the **temperature** of the acid affected the rate of the reaction?

b) What would you keep the same and what would you change?

The effect of catalysts

Catalysts are used to change the rate of some reactions. Catalysts are added to the reaction and are still there at the end of the reaction. **The catalyst itself is not used up during the reaction**.

Catalysts are usually added to **speed up a reaction**. They are very important in industry because they make reactions take place quickly at lower temperatures and pressures.

Negative catalysts slow down reactions. The rusting of radiators is slowed down by adding a negative catalyst to the water in central heating systems.

Industrial process	Catalyst
Haber Process for making ammonia	Pea-sized lumps of iron
Contact Process for making sulphuric acid	Vanadium(V) oxide
Making nitric acid	Platinum

We don't always know exactly how catalysts work. Finding a catalyst for a particular reaction has often been done by trial and error. Transition metals or their compounds are used as catalysts in many cases.

Investigating a catalyst

Hydrogen peroxide is a liquid that breaks down very slowly at room temperature. Hydrogen peroxide breaks down into water and the gas oxygen.

hydrogen peroxide → water + oxygen

$$2H_2O_2(aq) \rightarrow 2H_2O(l) + O_2(g)$$

If we mix liquid detergent with hydrogen peroxide, the oxygen given off makes a foam in the detergent. One way of comparing the rate at which hydrogen peroxide breaks down is to compare the height of the foam after different catalysts are added. A mixture of hydrogen peroxide and detergent are added to each measuring cylinder. A spatula of a different metal oxide is then added to each cylinder. The height of the foam is measured after a set time.

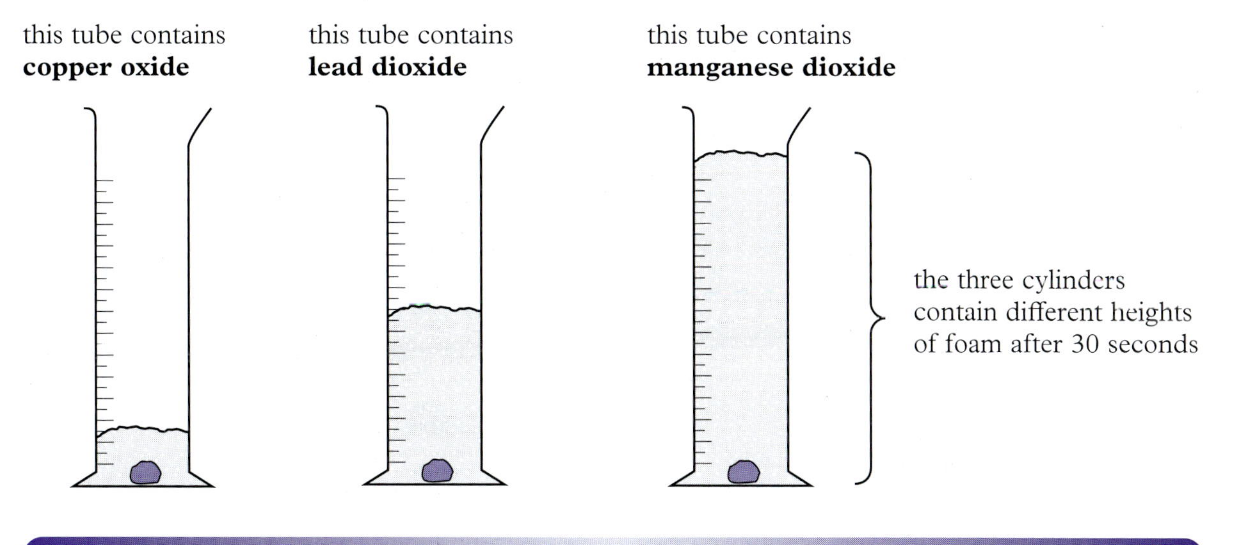
this tube contains
copper oxide
this tube contains
lead dioxide
this tube contains
manganese dioxide
the three cylinders contain different heights of foam after 30 seconds

Questions

1 What is a catalyst?

2 What is meant by a 'negative catalyst'?

3 Which metal oxide in the experiment above was the best catalyst?

Using the idea of particles to explain reaction rates

Chemical reactions

When two substances react together, their atoms bond together in different combinations.

these **bonds must break** **new bonds must form**

For this to happen:

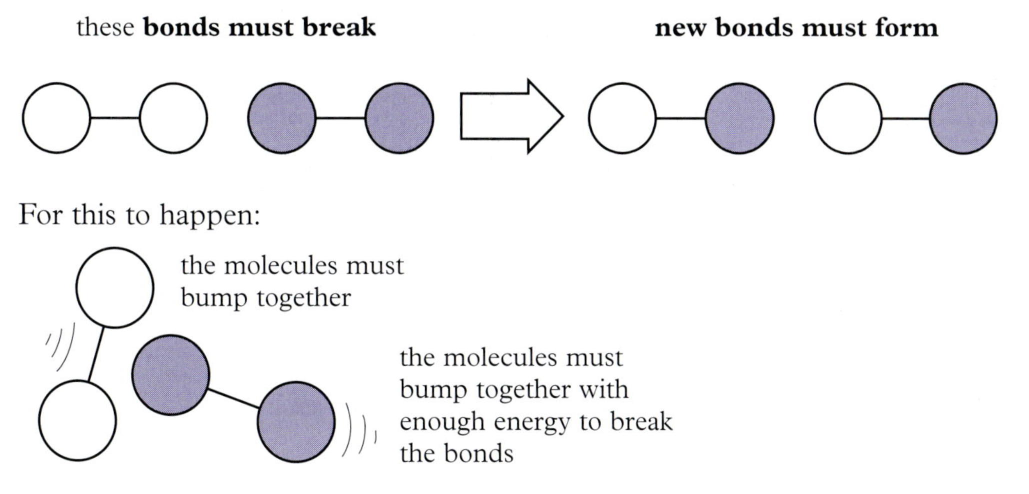
the molecules must bump together
the molecules must bump together with enough energy to break the bonds

This idea of how particles react together is called the **collision theory**.

Changing the reaction rate

If we want to make a reaction go faster, then we need to:

- get more particles to bump together, which means **more collisions**
- make them bump into each other **harder**
- make each bump more likely to **break bonds**

Getting more collisions

If we have **more particles** present in the same space, there will be more collisions because there will be more of them to bump into each other.

We can start with solid lumps and **increase their surface area**

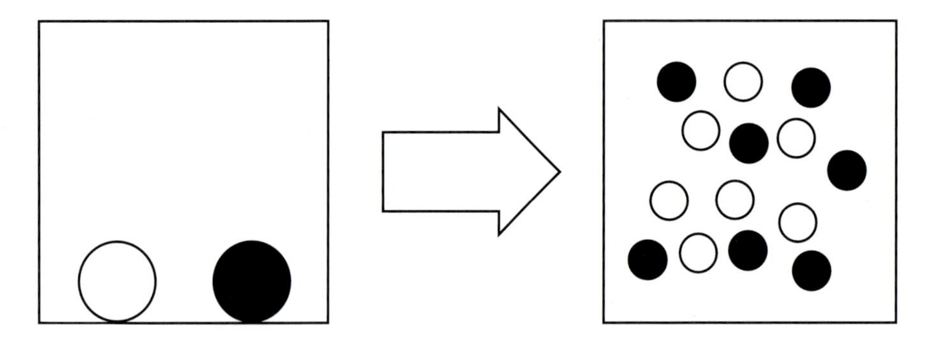

We can start with solutions and **increase their concentration**

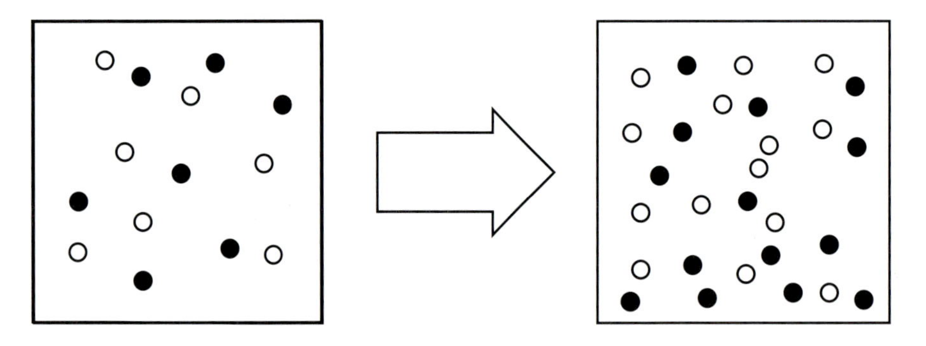

Questions

1 How would you increase the surface area of a lump of solid?

2 Two gases are squeezed together into a smaller volume. Their rate of reaction gets faster.

a) Draw before and after diagrams to help show how the number of collisions would go up.

b) Use different coloured circles for each gas. Remember to keep the same number of particles before and after the reaction.

Getting more energy into the collisions
If we increase the temperature of the reaction, then the particles move faster and they will **bump into each other more often** and with **more energy**.

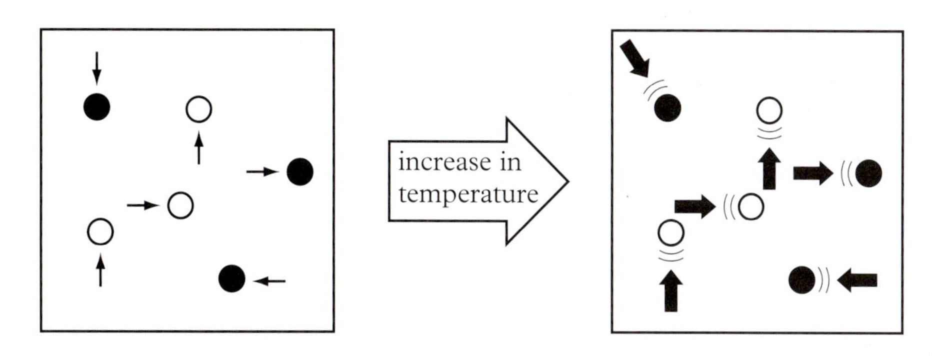
increase in temperature

Questions

1 a) What would happen to the rate if you cooled down a reaction?

b) Explain your answer in terms of particles colliding.

Making each bump more likely to break bonds
A catalyst works by making the collisions more likely to bring about bond breaking.

Enzymes

Enzymes are **biological catalysts**. They are proteins that are vital for all life. Enzymes control the speed of all the chemical reactions that go on in our bodies and in all other living things.

More and more enzymes are being used in industry too, because they are very effective catalysts. With them, reactions work faster at lower temperatures and pressures. For example, washing powders contain enzymes to help break down stains. Washing powders work well at low temperatures because they contain enzymes.

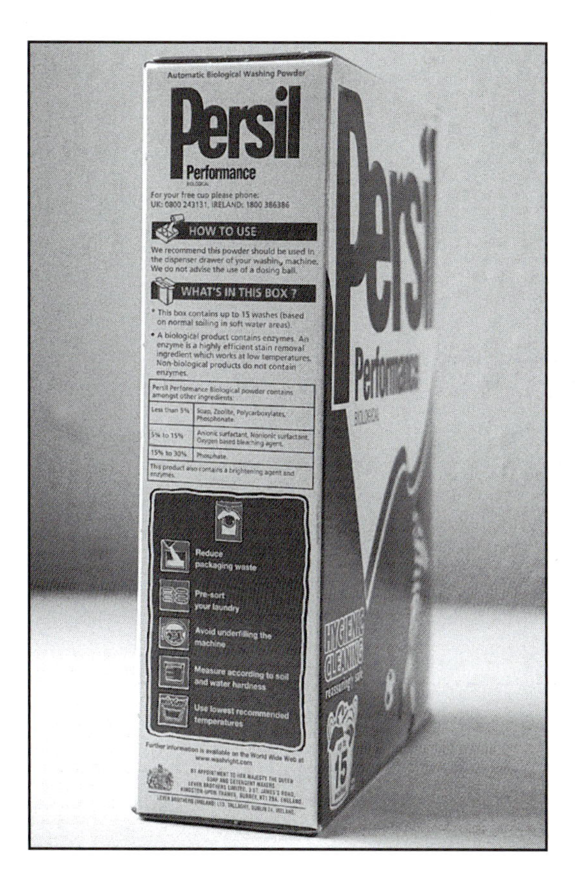
Automatic Biological Washing Powder
Persil
Performance
HOW TO USE
WHAT'S IN THIS BOX ?
Persil
Performance

For thousands of years, people have used micro-organisms that contain enzymes. These micro-organisms were used long before anything was known about enzymes. Micro-organisms are living things that are so tiny that they can only be seen with a microscope. Yeast is a micro-organism that contains the enzyme zymase. Yeast has been used for hundreds of years to make alcohol and to make bread rise.

Enzymes in yeast

Yeast uses sugars like glucose for energy. The energy is released by respiration. The yeast respires without using oxygen. This process is known as **anaerobic respiration**. The next equation is of anaerobic respiration:

$$\textbf{glucose} \xrightarrow{\textbf{zymase in yeast}} \textbf{carbon dioxide + ethanol + energy}$$

Both carbon dioxide and ethanol are useful by-products from the respiration of yeast. Ethanol is the chemical name of the alcohol in drinks.

In baking, the carbon dioxide gas bubbles into doughs so that bread dough rises and the bread is light and airy. The ethanol evaporates away.

In wine making, the ethanol is used. Yeast is added to a mixture of sugar, grapes and water, to make it gradually turn into ethanol. This process is called **fermentation**. Only about 15% of the solution ends up as ethanol. Yeast will stop growing at higher concentrations of ethanol than this.

Fermentation

Wine and beer are produced by fermentation on a large scale in industry. Fermentation can also be used to produce wine and beer at home with the equipment shown:

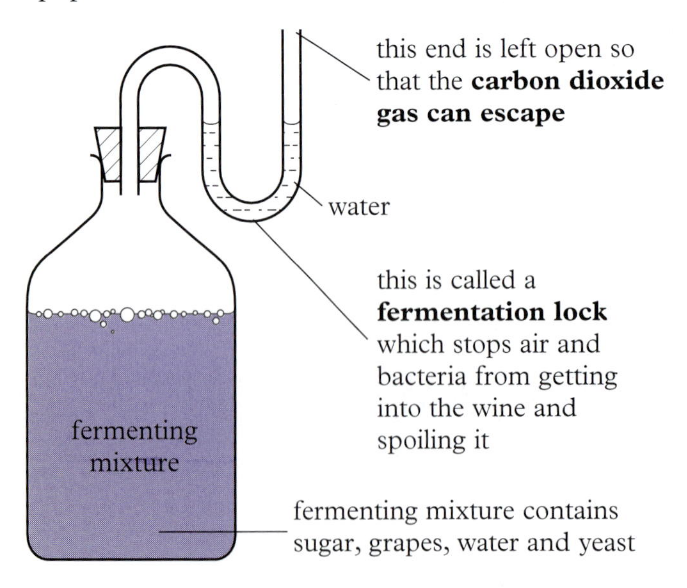
this end is left open so that the **carbon dioxide gas can escape**
water
this is called a **fermentation lock** which stops air and bacteria from getting into the wine and spoiling it
fermenting mixture
fermenting mixture contains sugar, grapes, water and yeast

Using enzymes produced by bacteria

Cheese and yoghurt are also made using enzymes in micro-organisms. In this case it is bacteria which produce the enzymes which act on milk.

There are a huge number of different enzymes and each one controls just one reaction.

How enzymes work

Enzymes work because of the shapes of their molecules. In this example there are two molecules which we want to react together.

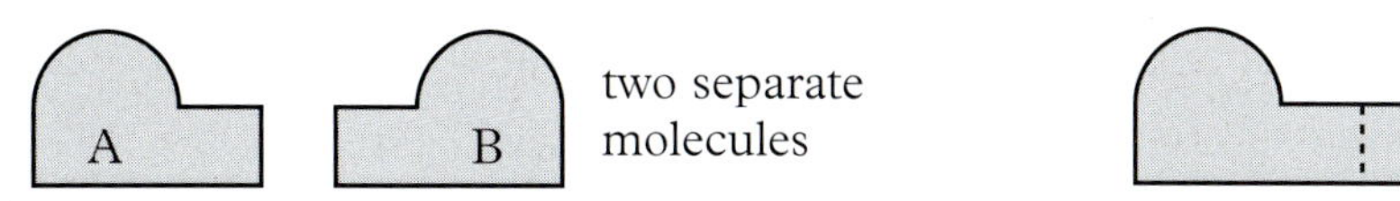
this is the reaction we want
A
B
two separate molecules
we want molecules A and B to react together like this

We can use an enzyme to make molecules A and B react together. The enzyme is exactly the right shape to line up the molecules A and B so that they react together.

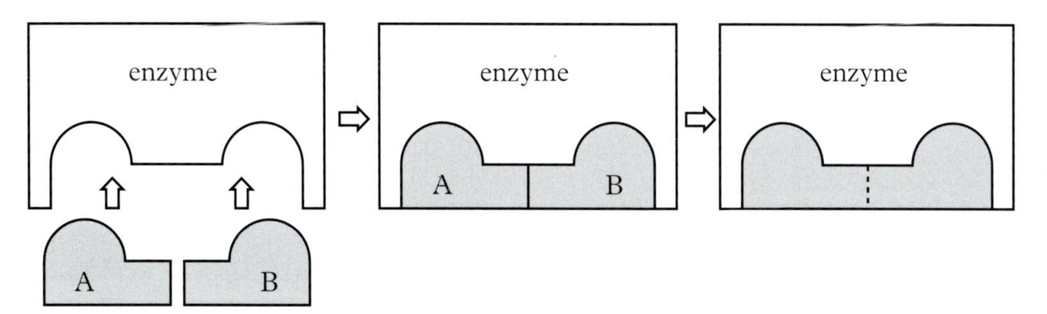
enzyme
A
B
enzyme
A
B
enzyme

Enzymes work best over a **small range of temperatures**. Often the **acidity must be just right** too. Excessive temperature or the wrong pH, affects the shape of the enzyme. We say the enzyme is **denatured** when its shape changes.

Many enzymes work best at about body temperature, 37 °C. This is called their **optimum temperature**. Above this, the enzyme is denatured, below this, the reaction goes too slowly because of the low temperature.

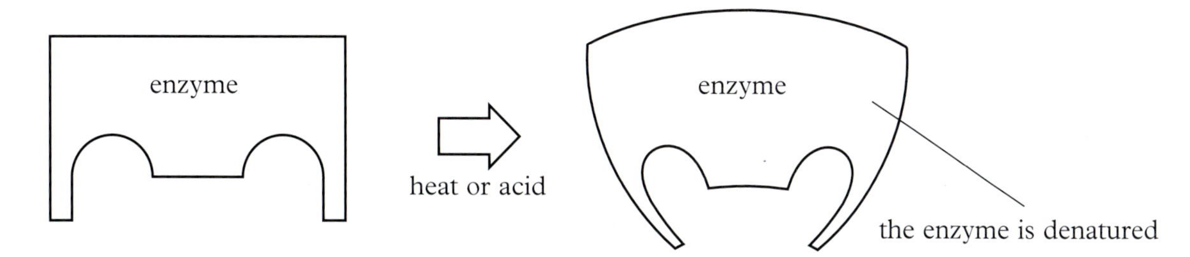
enzyme
heat or acid
enzyme
the enzyme is denatured

Questions

1. Food goes off because of chemical reactions which are often controlled by enzymes. Explain why food lasts longer in the refrigerator and even longer in the freezer.
2. Some food is preserved by pickling in vinegar, which is a household acid. Explain how vinegar could preserve food.
3. Explain why the shape of an enzyme matters.
4. What conditions cause enzymes to denature?

Summary

There are several things that change the rate of a reaction:

- Surface area of a solid – the greater the surface area, the faster the reaction.
- Concentration – the greater the concentration of a solution, the faster the reaction
- Temperature – the higher the temperature, the faster the reaction.
- Catalysts.

The collision theory states that the rate of a reaction depends on how many collisions there are between the particles in a reaction. The more collisions, the faster the reaction. The rate of a reaction also depends on how much energy there is in the collision itself (the more energy in the collisions, the faster the reaction).

Enzymes are biological catalysts found in living things. Enzymes are increasingly being used in industry because they make reactions go faster, or allow the reaction to go at the same speed but at a lower temperature or pressure.

Micro-organisms containing enzymes are used in the bread, beer and wine, cheese making and yoghurt industries.

Key words

catalyst	Something which changes the rate of a reaction without itself being used up.
concentration of a solution	The amount of dissolved material in a set volume of solution.
enzymes	Very efficient catalysts found in living things.
micro-organisms	Living material, such as bacteria and yeasts, that can only be seen with a microscope.
surface area	The area on the outside of a solid.

End of Chapter 16 questions

1 Zinc metal reacts with dilute hydrochloric acid to give off hydrogen gas.

$$Zn(s) + 2HCl(aq) \rightarrow ZnCl_2(aq) + H_2(g)$$

A lump of zinc was added to some hydrochloric acid to find out about the rate of the reaction. The results of the experiment were plotted on a graph. The shape of the graph is shown opposite:

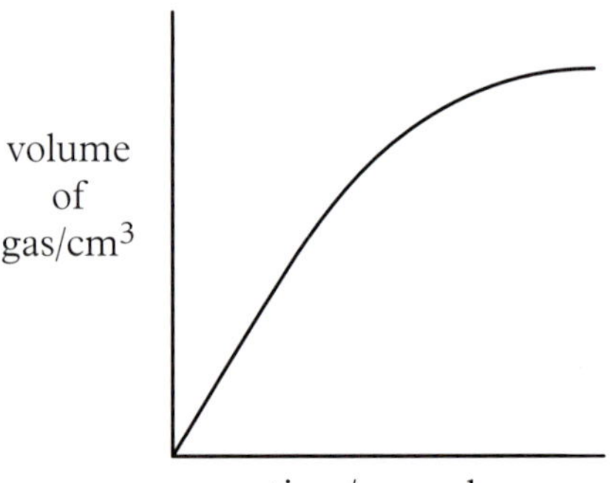
volume of gas/cm³
time/seconds

a) Write the word equation for this reaction.
b) Draw a diagram of the apparatus you might use to obtain the results shown on the previous graph.
c) What would happen to the rate of this reaction if powdered zinc were used instead of a lump of zinc?
d) Explain the effect of heating the hydrochloric acid before it is used.
e) Copy the previous graph and add to it a line that would show the effect of heating the acid.
f) Label each axis and add a title to the graph you have drawn in e).
g) If a more concentrated acid is used, the reaction goes more quickly. Explain this in terms of colliding particles. You may use a diagram to help.
h) Powdered copper acts as a catalyst to this reaction. Explain what is meant by a catalyst.
i) What is the test for hydrogen (see page 184)? Give the result you would expect if hydrogen is present.

2 The graph below shows the volume of carbon dioxide given off when a lump of calcium carbonate reacted with dilute hydrochloric acid.

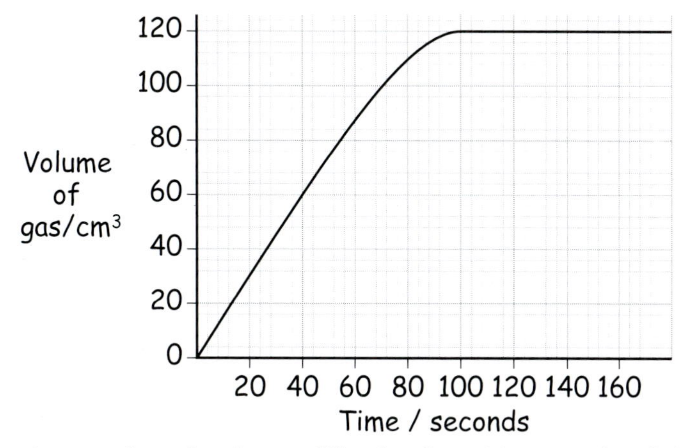
120
100
80
Volume of gas/cm^3
60
40
20
0
20 40 60 80 100 120 140 160
Time / seconds

a) What volume of gas is given off in the first 40 seconds of the reaction?
b) When was the reaction fastest? Explain your answer.
c) After how long did the reaction stop? Explain how this happened.
d) Give two ways in which this reaction could be slowed down.
e) What is the test for carbon dioxide (see page 184)? Give the result you would expect if carbon dioxide is present.

3 Hydrogen peroxide decomposes into water and oxygen at a very slow rate. The reaction can be speeded up using manganese dioxide as a catalyst. Manganese dioxide comes in the form of a black solid or as a powder. An experiment was done to compare the lump form with the powder form of manganese dioxide.

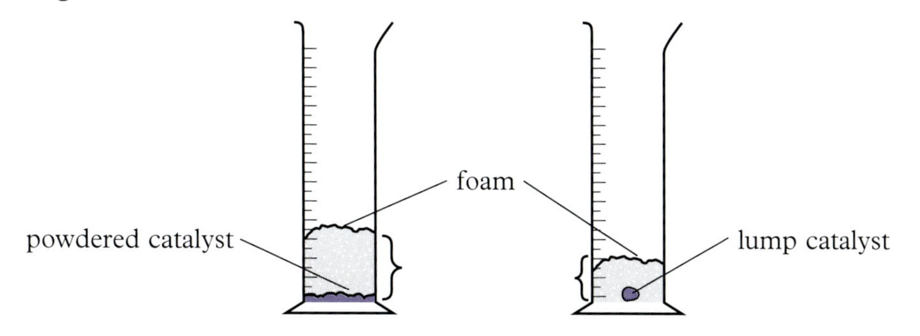
foam
powdered catalyst
lump catalyst

The detergent makes a foam which rises up the cylinder. The volume of the foam was measured every 20 seconds and the results were used to draw a graph.

Time / sec	Volume of foam / cm^3	
	powdered catalyst	lumps of catalyst
0	0	0
20	15	7
40	35	16
60	48	25
80	55	30
100	60	40

a) Plot the two sets of results on the same graph. Label each line.
   i) Is the lump form or the powdered form of manganese dioxide a better catalyst?
   ii) Explain your answer in terms of the surface area of the solid.
b) Manganese dioxide doesn't dissolve during this experiment. Describe an experiment to show that the catalyst was not used up in the reaction.
d) Describe a test for oxygen. Give the result you would expect.

4 Copy and complete the passage by choosing a word from the list below to fill in each blank. You may use a word once, more than once or not at all.

carbon dioxide **fat** **bacteria**
energy **yeast** **oxygen** **sugars**

Fermentation is a chemical reaction in which s_______ are converted into alcohol and c_______ . The process is speeded up by enzymes which are found in y_______ . Fermentation is an anaerobic process which means that it takes place without o_______ . In home brewing, the fermentation takes place in containers which are fitted with a fermentation lock. This allows the gas produced in the reaction to escape but keeps out b_______ which may be carried in by the vinegar fly.

17 The energy of reactions

When we burn a piece of wood or a lump of coal, we can feel the energy they give out in the form of heat. We use natural gas (methane) as a source of heat for gas cookers and gas fires. All fuels give out heat when they react with oxygen.

We call reactions which give out heat **exothermic**. Not all reactions give out heat. Whether they do or not depends on the bonds that hold the atoms together.

Breaking and making bonds

Whenever a chemical reaction takes place, atoms combine together differently. Whenever you light a Bunsen burner, methane reacts with oxygen to give carbon dioxide and water.

methane + oxygen → carbon dioxide + water

$$CH_4(g) + 2O_2(g) \rightarrow CO_2(g) + 2H_2O(l)$$

For this to happen, some bonds are broken and others are made.

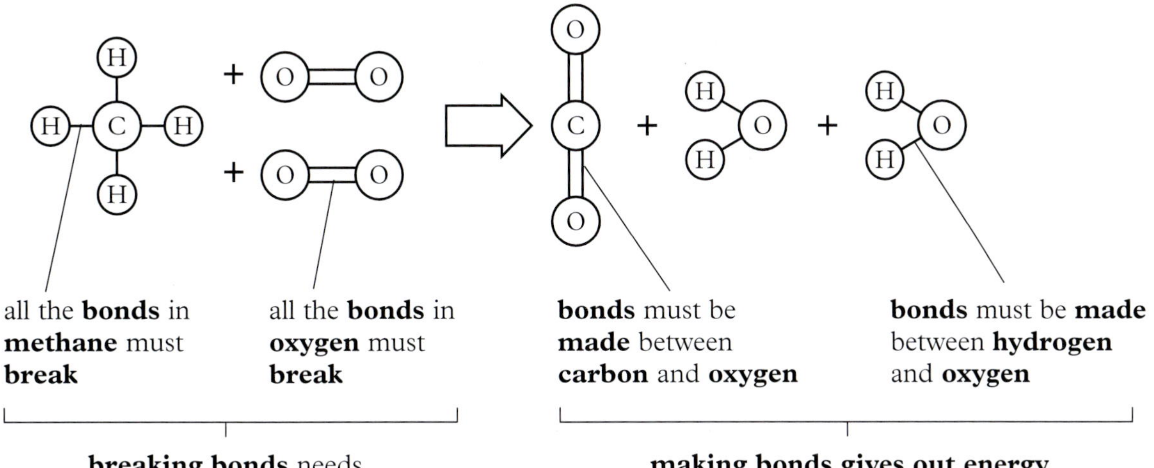

H
H
C
H
H
+
O
O
+
O
O
O
C
O
+
H
O
H
+
H
O
H
all the **bonds** in **methane** must **break**
all the **bonds** in **oxygen** must **break**
bonds must be **made** between **carbon** and **oxygen**
bonds must be **made** between **hydrogen** and **oxygen**
breaking bonds needs **energy** to be **put in**
making bonds gives out energy

Exothermic reactions

If more energy is given out than taken in, then the reaction is **exothermic**. The surroundings get hotter. It is easy to remember this term if you think of the word **exit** as meaning **out**.)

When a reactive metal reacts with an acid, the temperature of the acid goes up. This is an exothermic reaction.

We can use an energy diagram to show what is happening in exothermic reactions. The process begins with the reactants and ends with the products.

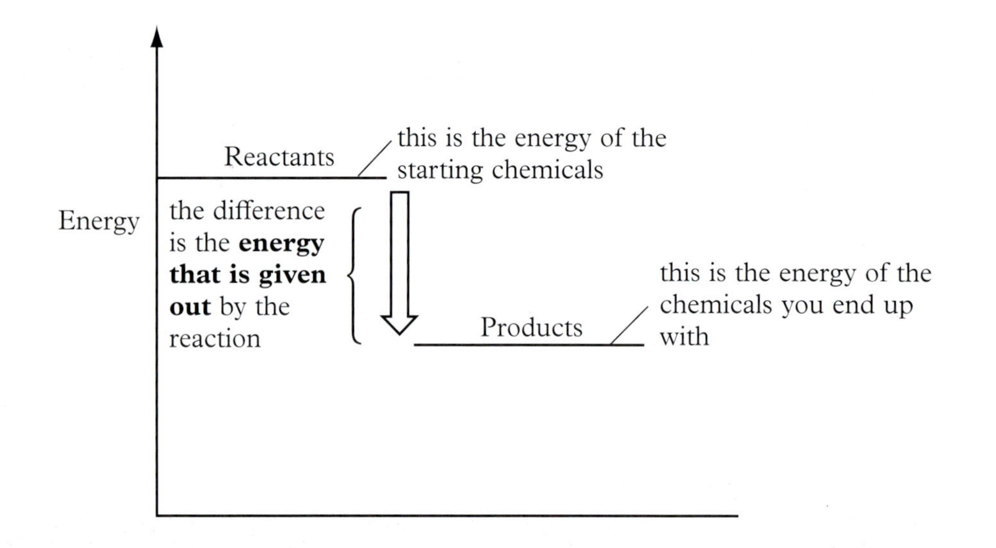
Reactants
this is the energy of the starting chemicals
Energy
the difference is the energy that is given out by the reaction
Products
this is the energy of the chemicals you end up with

Endothermic reactions

Reactions that are the opposite of exothermic reactions are called **endothermic** reactions. More energy is taken in than is given out. The surroundings will become cooler. We can also show this on an energy diagram.

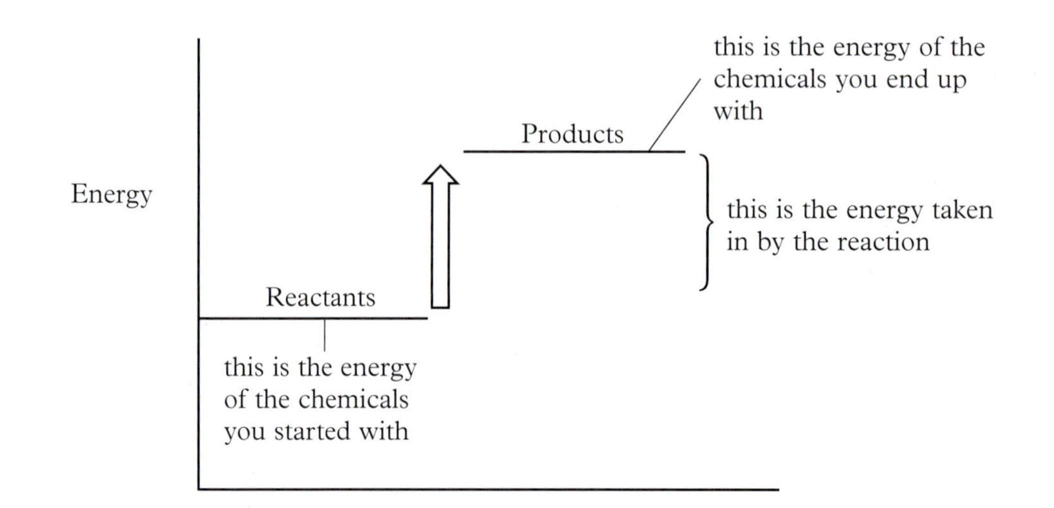
this is the energy of the chemicals you end up with
Products
Energy
this is the energy taken in by the reaction
Reactants
this is the energy of the chemicals you started with

Summary

Fuels give out heat when they burn in oxygen. Reactions that give out heat are called exothermic. Reactions that take in heat are called endothermic. Energy must be put in to break chemical bonds. Energy is given out when chemical bonds are formed.

The energy given out or taken in during a chemical reaction is the difference between the energy put in to break bonds and the energy given out when new bonds are made.

Key words

endothermic Reactions that take in heat.

exothermic Reactions that give out heat.

End of Chapter 17 questions

1 What term is used to describe reactions which give out heat?

2 Which word would you use to describe a reaction which takes in heat?

3 The table below shows three different chemical reactions.

Reaction	Temperature before the reaction / °C	Temperature after the reaction / °C
magnesium plus dilute acid	20	25
citric acid plus sodium hydrogencarbonate	18	14
calcium oxide plus water	19	28

a) What was the temperature change in the reaction between magnesium and dilute acid?

b) i) Which reaction gave out most heat?

ii) What would you feel if you held the test tube containing this reaction in your hand?

c) i) Which reaction took in heat?

ii) What would you feel if you held the test tube containing this reaction in your hand?

d) Draw an energy diagram for:

i) the reaction between magnesium and dilute acid

ii) the reaction between citric acid and sodium hydrogencarbonate.

4 Plants make their food by using the Sun's energy. What term would you use to describe this reaction?

Appendix

Tests

The following tests can be used to identify some common substances. They are worth learning by heart.

Test for	Test	Result
Hydrogen	Place a lighted splint in a test tube of the gas	Squeaky pop
Oxygen	Place a glowing splint in a test tube of the gas	Splint relights
Carbon dioxide	Add a few drops of limewater (calcium hydroxide solution) to a test tube of the gas	The solution goes milky
Chloride salts	Add a few drops of silver nitrate solution to a solution of the salt	A white precipitate is formed
Bromide salts	Add a few drops of silver nitrate solution to a solution of the salt	A cream precipitate is formed
Iodide salts	Add a few drops of silver nitrate solution to a solution of the salt	A pale yellow precipitate is formed
Water *	Add the liquid to a little white, anhydrous copper sulphate	The solid turns blue
Water *	Place blue cobalt chloride paper in the liquid	The paper turns pink

* Either of these tests can be used to confirm that a liquid is water. Both reactions are reversible:

- if the blue solid (called hydrated copper sulphate) is heated, it turns back to the white, anhydrous form
- if pink cobalt chloride paper is heated, it turns back to blue

The Periodic Table

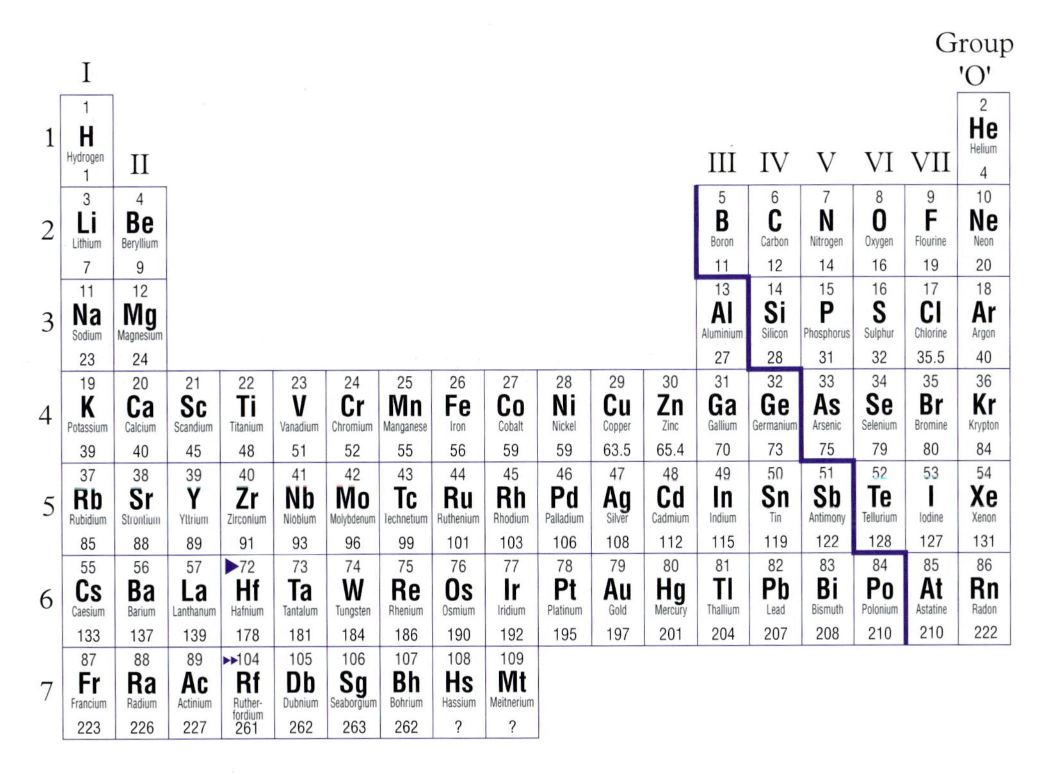

Period	I	II											III	IV	V	VI	VII	Group 'O'
1	1 H Hydrogen 1																	2 He Helium 4
2	3 Li Lithium 7	4 Be Beryllium 9											5 B Boron 11	6 C Carbon 12	7 N Nitrogen 14	8 O Oxygen 16	9 F Flourine 19	10 Ne Neon 20
3	11 Na Sodium 23	12 Mg Magnesium 24											13 Al Aluminium 27	14 Si Silicon 28	15 P Phosphorus 31	16 S Sulphur 32	17 Cl Chlorine 35.5	18 Ar Argon 40
4	19 K Potassium 39	20 Ca Calcium 40	21 Sc Scandium 45	22 Ti Titanium 48	23 V Vanadium 51	24 Cr Chromium 52	25 Mn Manganese 55	26 Fe Iron 56	27 Co Cobalt 59	28 Ni Nickel 59	29 Cu Copper 63.5	30 Zn Zinc 65.4	31 Ga Gallium 70	32 Ge Germanium 73	33 As Arsenic 75	34 Se Selenium 79	35 Br Bromine 80	36 Kr Krypton 84
5	37 Rb Rubidium 85	38 Sr Strontium 88	39 Y Yttrium 89	40 Zr Zirconium 91	41 Nb Niobium 93	42 Mo Molybdenum 96	43 Tc Technetium 99	44 Ru Ruthenium 101	45 Rh Rhodium 103	46 Pd Palladium 106	47 Ag Silver 108	48 Cd Cadmium 112	49 In Indium 115	50 Sn Tin 119	51 Sb Antimony 122	52 Te Tellurium 128	53 I Iodine 127	54 Xe Xenon 131
6	55 Cs Caesium 133	56 Ba Barium 137	57 La Lanthanum 139	▶72 Hf Hafnium 178	73 Ta Tantalum 181	74 W Tungsten 184	75 Re Rhenium 186	76 Os Osmium 190	77 Ir Iridium 192	78 Pt Platinum 195	79 Au Gold 197	80 Hg Mercury 201	81 Tl Thallium 204	82 Pb Lead 207	83 Bi Bismuth 208	84 Po Polonium 210	85 At Astatine 210	86 Rn Radon 222
7	87 Fr Francium 223	88 Ra Radium 226	89 Ac Actinium 227	▸▸104 Rf Rutherfordium 261	105 Db Dubnium 262	106 Sg Seaborgium 263	107 Bh Bohrium 262	108 Hs Hassium ?	109 Mt Meitnerium ?									

▶	58 Ce Cerium 140	59 Pr Praseodymium 141	60 Nd Neodymium 144	61 Pm Promethium 147	62 Sm Samarium 150	63 Eu Europium 152	64 Gd Gadolinium 157	65 Tb Terbium 159	66 Dy Dysprosium 163	67 Ho Holmium 165	68 Er Erbium 167	69 Tm Thulium 169	70 Yb Ytterbium 173	71 Lu Lutetium 175
▸▸	90 Th Thorium 232	91 Pa Protactinium 231	92 U Uranium 238	93 Np Neptunium 237	94 Pu Plutonium 242	95 Am Amercium 243	96 Cm Curium 247	97 Bk Berkelium 245	98 Cf Californium 251	99 Es Einsteinium 254	100 Fm Fermium 253	101 Md Mendelevium 256	102 No Nobelium 254	103 Lr Lawrencium 257

Index